THE
LIFE
MACHINES

DARIA MOCHLY-ROSEN, PhD, and EMANUEL ROSEN

THE
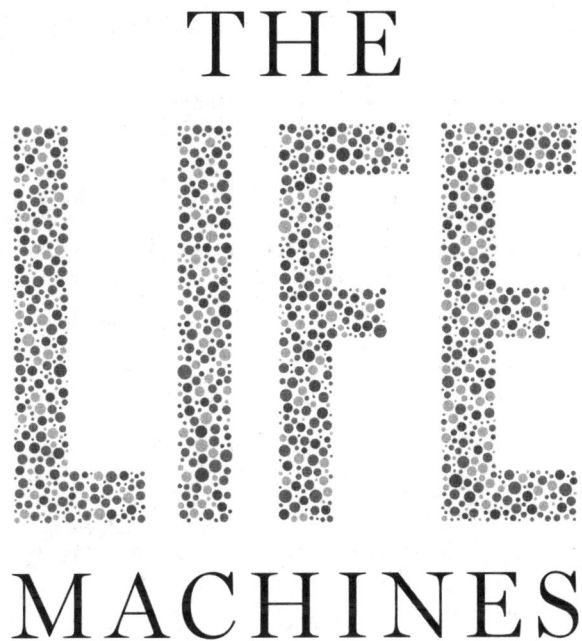
MACHINES

How Taking Care of Your Mitochondria Can Transform Your Health

SIMON ELEMENT

New York • Amsterdam/Antwerp • London
Toronto • Sydney/Melbourne • New Delhi

SIMON
ELEMENT

An Imprint of Simon & Schuster, LLC
1230 Avenue of the Americas
New York, NY 10020

To our family

CONTENTS

What Are Mitochondria and Why Should You Care?

In July 1989, a young mother named Patricia Stallings found her three-month-old baby, Ryan, listless in his crib. She rushed him to a hospital in St. Louis, Missouri, where the doctors ordered blood tests to figure out what was wrong. The lab report indicated the presence of high levels of ethylene glycol, a highly toxic compound found in antifreeze, so the suspicion arose that Ryan had been poisoned, and the authorities were notified. Luckily, Ryan's condition improved, but when he was released from the hospital, the Department of Social Services placed him in a foster home. Patricia and her husband, David, were allowed to visit Ryan once a week under supervision. Five weeks passed, and Ryan did fine, but after their sixth visit, Ryan suffered a severe attack of vomiting and was hospitalized again. The suspicions grew when investigators found out that during her last visit, Patricia was briefly left alone with the baby. Blood was drawn again, and the lab reported an even higher level of ethylene glycol in Ryan's blood. Patricia was arrested and charged with assault, and Ryan was put on a life support system. A few days later, he died in his father's arms. Patricia was charged with first-degree murder.

Three months later, while still in jail awaiting her trial for poisoning her son, something unexpected happened that would change everything: Patricia found out that she was pregnant again. In February 1990, she gave birth to another baby, David Jr. (nicknamed "DJ" by his parents), who was immediately taken from her and placed in a foster home. A few weeks after he was born,

DJ started to display similar symptoms to Ryan's symptoms—vomiting and breathing problems. He was brought by the foster family to a different hospital, blood was drawn, and when the lab report came back, the doctor's diagnosis indicated that DJ had an inherited disease—methylmalonic acidemia (MMA). This is a rare disease in which the mitochondria cannot break down certain metabolites. With this diagnosis, the doctors were able to treat him with a special low-protein diet that kept him alive. Importantly, this raised the possibility that Ryan, too, had MMA before he passed, and that Patricia Stallings was innocent. In fact, tests conducted on remaining blood samples from Ryan suggested that indeed, he *did* have MMA, but the defense didn't bring an expert witness to support this notion. Patricia's trial opened in late 1990 and the prosecutors were seeking a life sentence without parole. As we'll see in a moment, it was the involvement of two scientists that changed the course of the case.

In 2022, when we heard this story some thirty years after the fact, we were touched by what this young family went through. We were several months into writing this book, and this story demonstrated so powerfully why we embarked on this project—to make people aware of the key role mitochondria play in life and why it's so important to understand what they are really all about. Daria has been studying mitochondria for over two decades, and she has noticed a deep knowledge gap about mitochondria in the public and even in the medical community (as this case will demonstrate). In recent years, there has been an explosion of research in the field, which has widened the gap further, leaving plenty of room for confusion and false information. As a researcher studying mitochondrial dysfunctions in neurodegenerative and other diseases, Daria faces misconceptions about this topic every day. Emanuel is a writer, and as he became interested in these tiny organelles, he experienced firsthand how difficult it is for nonscientists to truly comprehend mitochondria's role in our lives. Together, we wanted to write a book that would be both accessible and scientifically accurate. The Stallings family story illustrated to us the price of mitochondrial illiteracy; it also reminded us that in the end, this book shouldn't be only about those organelles, but also about the humans who host them.

What Are Mitochondria and Why Should You Care?

Mitochondria aren't important only for babies like Ryan or his brother, DJ, but they play a pivotal role for every one of us at any age. These are tiny organelles that are present in almost every cell of our body. They transfer the energy from the food we eat into energy that fuels every movement, every thought, and every heartbeat. What are organelles? In the same way that we have organs like the heart, lungs, and brain that work in concert to make our body function, our cells have organelles that make the cell work. The most famous organelle in our cells is probably the nucleus, but when it comes to energy, it's all about the mitochondria.

Because of their critical role in transferring energy from the food we eat into ATP—a molecule that stores energy in the cell—mitochondria are popularly referred to as the powerhouse of the cell, but this term originated decades ago when scientists didn't know much about everything *else* that mitochondria do. Yes, the mitochondria are essential for energy transfer, but they also play a major role in several key cell functions. For example, they are responsible for cleaning your cells from toxic metabolites, triggering the recycling of some broken parts of the cell, detoxifying pollutants, replenishing building blocks of the cell, killing irreversibly damaged cells, fighting viruses, and sending signals about the status of our body. This is why we call them the "life machines"— because our life and death are in the hands of these tiny nanomachines. When they function perfectly, we're alive and healthy. When they malfunction, life is not the same, and it can even end. So knowing more about those organelles and knowing how to take care of them can affect your health as you go through life. We can't change our genetic makeup, but by tending to our mitochondria, we can reduce the burden on the cell and allow it to heal itself and even other cells. Simply put: How you live your life can either help your mitochondria or hurt them.

We divided the book into two parts so that you will first learn how mitochondria affect your health before we discuss how you can help them do an even better job. **In part I**, we'll cover the science of mitochondria—why these tiny organelles are the key to protecting your health—and why it is important to support them. In chapter 1, we'll introduce all the wonderful things that mitochondria do in addition to transferring energy from the food we eat. Chapter 2 will show that contrary to the view of mitochondria as stationary organelles, mitochondria are extremely dynamic. In chapter 3, you'll get to understand how the life machines falter as we age. In chapter 4, we'll explain how dysfunctional mitochondria are linked to common (and some less common) diseases, including neurodegenerative diseases, heart diseases, type 2 diabetes, cancer, psychiatric disorders, chronic fatigue syndrome, long Covid, autism spectrum disorder, infertility, and primary mitochondrial diseases. In the final chapter of this part—chapter 5—we'll discuss cutting-edge medical interventions that involve mitochondria and can transform your health and the way we think about medicine. Understanding how these medical interventions work will also deepen your understanding of the advantages you gain by making healthy lifestyle choices.

In part II, we'll cover practical lifestyle changes that you can make to be good to your mitochondria. In essence, we'll show how by taking care of your mitochondria, *you* can transform your health. Our goal is to give you a deeper understanding of recent scientific discoveries related to mitochondria so that you can evaluate new information, separate fact from fiction, and make educated choices about your health. In chapter 6, we'll talk about the importance of exercise, which most scientists believe is good for your mitochondria. In chapter 7, we'll review some of the latest research on nutrition, fasting, vitamins, and supplements. Chapter 8 will cover the relationships between mitochondria and the bacteria that live in your gut (and are part of your gut microbiome). In chapter 9, we'll discuss the importance of proper sleep. Chapter 10 will focus on the negative effects of prolonged emotional stress on our mitochondria, and we'll discuss some stress-reduction practices. In chapter 11, we'll focus on things to avoid, including (but not limited to) smoking, vaping, alcohol, air pollution, exhaust fumes, and pesticides. Finally, in the epilogue, we'll share some thoughts about the future of medicine and reiterate some habits you can adopt to help your own life machines thrive.

Ten Quick Facts About Mitochondria

To help readers who have never heard about mitochondria (or forgot what they knew), here are ten quick facts about them. Don't worry about remembering every single one of these because we'll revisit them throughout the book:

1. **They transfer energy.** Mitochondria transfer energy from the food we eat to small molecules called adenosine triphosphate (ATP), which are then used to power the cell's activity. Mitochondria accomplish this task using the oxygen we breathe. Notice that we didn't say that they *create* energy—instead, they unpack the energy locked in our food. Their two main fuel sources are fatty acids (which come from breaking down fat) and glucose (which comes from breaking down carbohydrates). What about proteins? Proteins, and specifically their building blocks that are called amino acids, can also be used as a fuel source, especially when fatty acids and glucose levels are low.

2. **They have two membranes.** One inner, one outer. ATP is produced through a multistep (and quite remarkable) process that involves the inner membrane. Some specialized mitochondria, rather than producing ATP, utilize energy to produce heat. Yes, mitochondria also keep us warm.

3. **They are the ultimate multitaskers.** Some people view mitochondria as batteries that provide energy, but as mentioned, they do so much more. Here is a partial list: They send signals that

regulate cell activities; they kill cells that should die; they help the cell fight viruses; and they produce essential building blocks of the cell.

4. **They move around.** Biology students are accustomed to seeing a static cartoon of a bean-shaped organelle, but mitochondria are extremely dynamic: They change their shape, split and merge, move within the cell, send stuff to other cells, and even travel from cell to cell within our body.

5. **Their functions deteriorate over time.** Like everything else in our body, as we age (which starts decades before we retire), their functions naturally deteriorate, so our mitochondria need all the help we can give them.

6. **There are about 10 million billion mitochondria in your body.**[1] The single form of mitochondria is a mitochondrion, but these organelles always exist in groups. In a typical diagram of a cell, you'll find three or four mitochondria, but in reality, some cells have hundreds, while others have thousands. Their number in a cell varies based on the need for energy, so heart cells and neurons, for example, have the highest number of mitochondria. The only cells in our body that don't have mitochondria are red blood cells (possibly because they need all the room to carry oxygen).

7. **They are believed to originate from ancient bacteria.** The prevailing theory today is that mitochondria originated from bacteria that were engulfed by early cells about 1.5 billion years ago. This was the beginning of a beautiful friendship that enabled all complex life forms on planet Earth. (Yes, animals and plants have mitochondria, too.)

8. **Mitochondria have their own DNA.** Related to their origin in ancient bacteria, mitochondria have their small DNA (mtDNA), which we inherit only from our mothers. mtDNA encodes thirteen important proteins in the mitochondria, which gives them control over critical components essential for their operation.

9. **Mitochondria work closely with nuclear DNA.** The fact that mitochondria have their own DNA doesn't mean that they are autonomous. Each mitochondrion contains an estimated eleven hundred to fourteen hundred distinct proteins, and all but thirteen of them are encoded by nuclear DNA (which means that their assembly instructions are stored in the nucleus).

10. **Many chronic diseases are associated with mitochondrial dysfunctions.** These include age-related diseases, like diabetes and dementia, and also primary mitochondrial diseases, which are rare, inherited conditions that we'll discuss later.

Why Knowing Makes a Difference

Speaking of rare inherited diseases, MMA is one of them, which brings us back to Patricia Stallings's story. Following the revelation that Patricia's newborn son had MMA, the newspapers in St. Louis covered this story closely. William Sly was the chairman of the biochemistry department at Saint Louis University, and he remembers that one young faculty member at his department was particularly troubled by the case. James Shoemaker had just joined Sly's department, and Sly encouraged him to start a lab to screen for metabolic diseases. As mentioned, there were some remaining blood samples from Ryan (Patricia's firstborn child who died), and Shoemaker examined those samples. He was the one who found evidence that Ryan also had MMA.

What Shoemaker found in Ryan's blood was a substance called propionic acid. As expert scientists in the field, he and Sly knew this acid could build up in the body of MMA patients and could easily be mistaken for ethylene glycol. But this knowledge wasn't widely disseminated in the medical community. On January 31, 1991, Patricia Stallings was found guilty, and a month later she was sentenced to life in prison without parole.

Following her sentencing, Jim Shoemaker was restless. He and another young faculty member, Joe Hoffman, were obsessed with this story, and every day Sly would see those guys outside his door talking about the case. One day, Shoemaker walked into Sly's office and told him that the case would be featured on the TV show *Unsolved Mysteries*.

"That evening, I watched the show with my wife, and it was heartbreaking," Sly told us. When the program ended with Patrica's sentencing, Sly knew he had to do something about it.

To see if other labs would misread propionic acid as ethylene glycol, Sly and Shoemaker sent blood samples spiked with propionic acid to seven different laboratories. Sure enough, three out of the seven labs misidentified the propionic acid and reported it as ethylene glycol. Sly sent a letter to the county prosecutor George McElroy with this information and urged him to revisit the case. McElroy was impressed by Sly's letter, but he wanted a second opinion. They found Piero Rinaldo at Yale who received all the raw material from the labs and ran the test again. Rinaldo confirmed their suspicions: Ryan was not poisoned; he had MMA.

On September 20, 1991, in a room packed with journalists and TV crews, Jefferson County prosecutor George McElroy announced that the State of Missouri was dismissing all pending charges against Patricia Stallings. He apologized to the Stallings, who were sitting next to him, both personally and on behalf of the State of Missouri. During that meeting, it was also announced that eighteen-month-old DJ would be released to his parents. Patricia gasped, smiled widely, and fought back tears. Her husband, David, covered his eyes with his hand and lowered his head.

Patricia Stallings wasn't the killer. The killer was the disease—methylmalonic acidemia. The poison wasn't antifreeze but propionic acid, which accumulates when a vital chemical reaction in the mitochondria goes wrong.[2] The lab error in this case was driven by lack of awareness of this process. If Ryan had been diagnosed with MMA, the doctors could have treated him with the same special low-protein diet that kept DJ alive. The lesson we can learn from this story is universal: Because various mitochondrial dysfunctions can affect all of us at different times of our lives, it is so vital that we are all aware of the importance of mitochondria and know how to tend to them. Mitochondria are complex, no doubt. But if we want to understand health and disease, we need to learn more about those misunderstood organelles—the life machines.

Since there's so much new research on mitochondria, and because they touch every aspect of our lives, we realized very quickly that we would need the help of experts in many subfields. Also, while we're both health-conscious individuals, we cannot be described in any way as health gurus. So we talked to dozens of scientists, delved into numerous research papers, and sat through many conference lectures. We met patients and their families who deal with severe cases of mitochondrial dysfunctions, and we were moved by their courage and determination. We spoke with researchers who were excited about these tiny organelles and the potential of their research to change people's lives. Scientists are sometimes stereotyped as living in ivory towers, but we often met caring people in lab coats who are determined to get to the truth to help other people. When we talked with William Sly, he told us that not long after the charges against her were dropped, Patricia Stallings showed up in his office one day. "She said: 'Can I give you a hug?'" Sly remembered, "and I said, 'Sure,' and that's the last time I saw her."

Our collaboration on this book started as we awaited the gift of a new life. We moved to Boston for a few months to help our daughter, who was expecting a new baby, and on long walks through freezing Boston, we started a little game: For five minutes, Daria would tell Emanuel something interesting about mitochondria, and then he would have to explain it back to her in his own words. It was a humbling experience for both of us—Emanuel, because without any science background, he could barely string together a coherent sentence; and Daria, because she realized she could hardly communicate without relying heavily on scientific terms. Yet gradually, we found a way to talk about it through metaphors and anecdotes. As we finally started speaking the same language, a new sense of awe bubbled up in us: awe for life and the tiny machines that enable it.

Back home in California as proud grandparents, we couldn't stop sharing not just adorable grandkids' photos, but also every fascinating fact from the latest mitochondria studies. We had never collaborated before—setting aside the project of raising four kids—but we *had* to write this book, and the division of labor was clear. As the scientist, Daria was the one who dug deep into scientific reports and asked detailed questions. As a writer, Emanuel was the one who asked simple questions, like "What does it mean to me?" and "Is there anything one can do to help our mitochondria?" Together, we've tried to write a book that is both clear and scientifically rigorous. We used the first-person plural throughout the book to simplify things, but of course, the scientific opinions are Daria's. Everything else reflects our common experience on this three-year quest.

We invite you to join our journey and then embark upon your own.

PART I

HOW YOUR MITOCHONDRIA HELP YOU

CHAPTER 1

Yes, the "Powerhouse," but So Much More

The other day, we saw a dead bird on our porch, lying on its back. It was small, brown, and motionless. We're always struck by the stillness, the absolute lack of movement, when we see a dead creature. Everything was still there—the beak, the eyes, the wings, the tail feathers, the tiny feet. So what was missing? Energy. Without energy, there is no life. That energy comes from tiny machines that are present in almost every cell of every plant and every animal on this planet— mitochondria. There is no life without mitochondria. They give us life; they can take it away, and when they malfunction, our health deteriorates. As you'll see in this book, it pays to keep them happy. Mitochondria are the primary fuel source for all the body's activities. Consider, for example, what is happening in your body as you are reading this paragraph. Muscle cells in your heart contract to pump blood through your arteries, other muscle cells expand your rib cage to pull air into your lungs, and specialized cells in your eyes send signals to the brain where neurons make sense of the text you're reading. All these activities require energy, and if you could zoom into each one of these cells, you would see your mitochondria working.

Each cell has hundreds, and some have thousands of mitochondria, and if you could further zoom into each one, you would see something remarkable: tiny rotors that are spinning at a rate of hundreds of revolutions per second, transferring energy from the food you ate to a small molecule called adenosine triphosphate (ATP), which is then used to power the cell's activity. This is not a metaphor. At this very moment, actual rotors are spinning in your cells enabling your body to do stuff. This was a partial (and very simplified) explanation of what happens in your body when you read a paragraph in a book. Just imagine what's happening (and how much energy is needed) when you go for a run, do yard work, or dance. And we haven't even begun to address the millions of cells in your brain, liver, stomach, pancreas, kidneys, and other organs that are busy with routine maintenance tasks. Mitochondria are the main fuel source for all these activities.

Yet as remarkable as the process of energy transfer is, it is just one of the tasks that mitochondria do, and only once you understand their wide range of functions will you be able to appreciate their full impact on your health. Anyone who has taken a high school biology class likely associates mitochondria with the catchphrase "the powerhouse of the cell." While there are good reasons for this strong association, it paints a very partial picture of what mitochondria do. In this chapter, we'll correct this major misconception by focusing on four of their many other functions. Mitochondria act as signaling stations that regulate cell activities, as assassins of cells that should die, as fighters that protect us from viruses and other invaders, and, perhaps most aptly, as bustling factories—metabolic hubs that produce the building blocks of life. What exactly do we mean by "metabolic hubs"? You can think of each mitochondrion as a factory where materials that came from your food are broken down, and where new materials are produced. To understand those busy centers of metabolic transformation, we should go back to an important discovery made in the first half of the twentieth century.

Mitochondria Are Metabolic Hubs and the Story of One Gold Medal

On June 19, 1933, a thirty-two-year-old man stood on the platform at the Freiburg train station in Germany holding a one-way ticket to Strasbourg, France. He was relatively short, with round glasses, somewhat nervous and

worried. Two months earlier, this man, Hans Krebs, had received a letter placing him on indefinite leave from his position as a lecturer at the local university
because he was Jewish. He boarded the eleven o'clock train and started his long
journey to his final destination—England.[1]

In England, he would some day make one of the most important discoveries about mitochondria, which would earn him the Nobel Prize. But now on
that train, it's not hard to imagine what was going through his mind. He had
just lost his job; he was fleeing his home country; and he wasn't sure what lay
ahead. A few days after arriving in England, Krebs met the renowned scientist Frederick Gowland Hopkins in Cambridge who admired Krebs's previous
work in biochemistry. Hopkins was apologetic that he could offer him only a
one-year position at a low salary, but the young man was happy to have a job
and accepted it on the spot. Three weeks later, Krebs received a letter from the
rector of his German university, the philosopher Martin Heidegger, who was
an enthusiastic supporter of Hitler, dismissing Krebs from his post. After a year
in Hopkins's lab, Krebs was offered a position at the University of Sheffield,
where he would spend the next nineteen years and where he made his most
important discovery—the Citric Acid cycle, which is also known as the *Krebs
cycle* (the term we'll use here).

If you were traumatized in high school or college by having to memorize
the Krebs cycle, you'll be happy to hear that we won't get into the nitty-gritty
details here. It is essentially the chain of chemical reactions that make up the
first step in the process of ATP production inside the mitochondria. You can
think of the Krebs cycle as a circular conveyor belt, not unlike the baggage carousel at the airport, where you pick up your suitcase after a flight. But instead
of suitcases, this carousel is loaded with molecules called acetyl-CoA that, after
going through some chemical reactions, come out as important compounds
that are essential to producing most of the ATP in the mitochondria. Acetyl-
CoA comes from the food you eat. Think of the delicious chicken pesto panini
you had for lunch: the carbs, fats, and proteins from your sandwich are broken
down in your digestive system into simple sugars (glucose), amino acids, and
fatty acids that travel through your blood to your cells. It is in your cells that

some of these materials are converted into acetyl-CoA and other intermediates of the Krebs cycle, which are loaded into this intricate baggage carousel. The two most important "suitcases" that get off that carousel are NADH and $FADH_2$, two compounds that are essential for ATP production. (More about them in chapter 3.)

Initially, Krebs's work didn't draw much attention, and his paper on the subject was rejected by the prestigious journal *Nature* in less than a week. (Years later, an editor at *Nature* would admit that it was the most egregious error in the journal's history.[2]) After some initial skepticism, more and more scientists recognized the significance of Krebs's work. He set the stage for scientists to further examine the critical role of mitochondria in metabolism. On October 22, 1953, twenty years after he had to leave Germany, Hans Krebs heard on the BBC news that he won the Nobel Prize. A couple of months later, at the Stockholm Concert Hall, he was handed the gold medal from the king of Sweden. Incidentally, Krebs never forgot the generosity of those who helped him as a refugee, and in 2015, his son found a way to pay it forward. He sold the Nobel Prize medal and, using the proceeds, established a trust to support young biomedical scientists who have been forced out of their country because of conflict, discrimination, or danger.

To reiterate what we've said so far about the Krebs cycle: It takes place inside the mitochondria, and it is the first step in the process of ATP production. Unfortunately, many discussions of the Krebs cycle stop right here. Yes, the Krebs cycle is justifiably seen as a critically important part of the energy transfer process, but here comes a big but. . . .

The Building Blocks of Life

But the Krebs cycle isn't only about energy transfer. If you think of your body as a building, the production of many of its building blocks—the fundamental elements of the cell—starts in the Krebs cycle and in other reactions within the mitochondria. To look at just one example of such a building block, let's turn again to what's happening in your body right now: neurotransmission—the release of chemicals at the end of a neuron to communicate with another neuron. Whether you're processing what you read in the last paragraph or thinking about your plans for the weekend, these activities involve lots of neuron-to-neuron communication. The most abundant neurotransmitter, the agent that communicates between neurons in your brain, is called glutamate, and its building block (called alpha-ketoglutarate, or AKG) is generated in the Krebs cycle. Too much glutamate in the brain can cause seizures and mental exhaustion, whereas too little can cause sleep problems and low energy.

Consider three important molecules in the cell: proteins, DNA, and lipids (fat). Most of the building blocks for these molecules are made or processed in mitochondria. Your DNA is made of building blocks called nucleotides and some of them are processed in your mitochondria. Some building blocks of proteins (called amino acids) are generated in your mitochondria, too, and the same is true for some building blocks of lipids. All these are the building blocks of life. Mitochondria don't produce all these building blocks alone—many cell functions are joint efforts of several organelles—but if you start looking at how parts and components of the cell are made, you're likely to repeatedly come across critical contributions by the mitochondria.

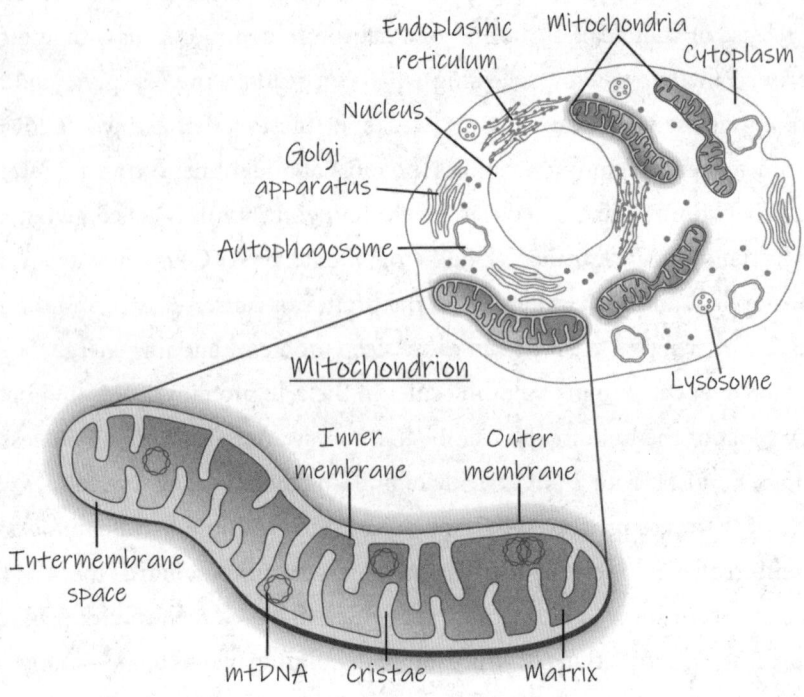

Endoplasmic reticulum
Mitochondria
Cytoplasm
Nucleus
Golgi apparatus
Autophagosome
Lysosome

Mitochondrion

Inner membrane
Outer membrane
Intermembrane space
mtDNA
Cristae
Matrix

A simplified illustration of a mitochondrion and a cell. The upper part of this illustration depicts a cell with some organelles. Many of the building blocks for the molecules that make the cell are processed in mitochondria. Real cells can have hundreds or even thousands of mitochondria. The lower part of this illustration depicts a mitochondrion. The space between a mitochondrion's outer and inner membrane is called the intermembrane space. The space in the middle of a mitochondrion is called the matrix, and this is where the Krebs cycle takes place. Note that the organelles are not drawn to scale.

For the next example of a building block, you may want to take a deep breath (which is a good idea anyway). How is breathing related to mitochondria? If you answered that mitochondria use the oxygen you breathe to convert the food you eat into ATP, you are correct, but only partially. Many people are not aware of the role that mitochondria play behind the scenes in delivering oxygen to your cells. Oxygen is delivered through a protein in your red blood cells called hemoglobin. The part of hemoglobin that enables capturing oxygen and delivering it to all your tissues is called heme, and its production starts and ends in the mitochondria. Heme also helps remove carbon dioxide from your cells and deliver it back to your lungs, which in turn remove it from your body when you exhale. If this isn't a big enough job, here's another way heme helps you: After you go to the gym, your body adds myoglobin to your muscles, which increases your ability to store even more oxygen in your skeletal muscles for the next time you exercise, and heme makes up the oxygen-binding component in myoglobin. We can go on and on with the list of compounds that are processed with the help of mitochondria, but for the sake of brevity, let's just mention three more: testosterone (male hormone), estrogen (female hormone), and cortisol (stress hormone).

WASTE PRODUCTS

As part of their role as metabolic hubs, mitochondria not only provide building blocks to your body, but they also participate in removing some waste products from the cell. We already mentioned carbon dioxide. This waste product that leaves our body when we exhale is produced in our mitochondria. Another waste product is urea, and its production starts in mitochondria. When our body breaks down proteins, it creates ammonia, which is highly toxic, so it is converted into urea, which is removed in our urine. Known as the urea cycle, this was another discovery of Hans Krebs, which he had made while still in Germany and which established his international reputation.

Although the Krebs cycle is usually mentioned in the context of energy, Hans Krebs was very aware of the cycle's dual role of providing energy *and* building blocks, and he referred to this point when he was awarded the Nobel Prize.[3] He also said that the research he'd been doing did not lead to the kind of knowledge that can be expected to give immediate practical benefits to mankind, but that he was convinced that it will eventually help solve some of the practical problems of medicine. Seventy years later, there is plenty of evidence of the impact of his discoveries on our understanding of human health. Thanks to Hans Krebs, we understand that mitochondria are amazingly efficient factories that provide building blocks that affect every aspect of our health.

Mitochondria Help Kill Cells That Should Die, and How This Shaped Your Fingers

In 1997, when the cell biologist Guido Kroemer finished a lecture about the role of mitochondria in programmed cell death, one of his esteemed colleagues stood up and exclaimed: "Mitochondria are bullshit!" Before we discuss the colleague's harsh reaction, let's explain what programmed cell death is. Imagine that you're the manager of a thriving chemical plant. Every morning, you walk around the plant to review which pipes or valves need to be lubricated, adjusted, or replaced. The materials that flow in the system are highly corrosive and toxic, and being the responsible manager that you are, you ensure that pipes and valves are replaced at the right time. As the years go by, more and more parts need to be replaced, and one day you get word from corporate headquarters that the plant is beyond repair and is going to be shut down. Do you simply lock the doors and abandon the place? Of course not. To avoid an environmental catastrophe, it must be done in an orderly fashion: Chemicals need to be removed safely with special attention to prevent any dangerous interactions among them.

Every cell in your body is like that chemical plant. During the lifetime of the cell, broken parts are replaced, and certain chemicals are kept away from one another to prevent unwanted interactions. When the cell is damaged beyond repair, it needs to be removed safely because the content of a cell can be highly toxic to other cells. This process is known as apoptosis—programmed cell death—and it originates and is orchestrated by the mitochondria.

For years, scientists assumed that programmed cell death was completely controlled by the nucleus. Some suspected that mitochondria were involved, and Guido Kroemer made significant contributions to establishing this involvement. But no one was able to prove it until 1996, when a young scientist named Xiaodong Wang at Emory University in Atlanta made a stunning discovery. He was trying to identify what activates caspase-3, an enzyme that was known to be involved in programmed cell death, and his experiments kept showing that it was a protein called cytochrome c. This didn't seem to make sense because it was well established that cytochrome c played a key role in the energy transfer process in the mitochondria. Is it possible that the very same protein that helped the cell to live also played a role in killing it? Wang told us that initially he dismissed his own findings, but cytochrome c kept showing up as the activator of caspase-3. Even when he had very clear results, he still hesitated to make his work public. To test the waters, he presented his results to two Nobel laureates, Michael Brown and Joseph Goldstein at the University of Texas (he had previously worked in their lab). Wang opened his presentation by saying that if he didn't manage to convince them, he would not submit the article, because he doubted that he would be able to convince anyone else. When he completed the presentation, Michael Brown left the room and started pacing the corridors. Goldstein stayed in the room, thinking. Forty minutes later Wang heard the verdict: He should submit the paper to the highly regarded journal *Cell*.

The idea that mitochondria are involved in programmed cell death faced a lot of skepticism within the scientific community, even after Wang's paper was published.[4] The harsh reaction that Guido Kroemer heard wasn't the only incident. Eventually, though, Wang and others ran additional experiments that proved the mitochondria's critical role in triggering programmed cell death. Today, it is unequivocally established that mitochondria play an integral role in this job. And it's a big job: In a single day, programmed cell death takes place in billions of cells in your body. When apoptosis doesn't take place, in other words, if the cell just dies without the safe removal of its parts, toxic chemicals leak into the space between cells, damaging or killing neighboring cells and triggering inflammation.

Apoptosis does more than protect us from inflammation and damage to neighboring cells. There are almost always potentially cancerous cells in our body. This is largely unavoidable since as we age, some cells collect too many mistakes in their nuclear DNA, causing them to divide uncontrollably. Luckily, our body has some safety mechanisms to fight cancerous cells. One of those mechanisms is programmed cell death that eliminates cells with potentially dangerous mutations. Sometimes the process of apoptosis is broken, so cells that we wish were dying keep multiplying. This is part of cancer. Many anti-cancer drugs work by reactivating the process of apoptosis and thus triggering the cancer cell to die.

Through apoptosis, mitochondria also participate in the beautiful process of shaping a human while in the womb. For example, the hands of the fetus in the early stages of development first look like mittens, and then a magical process begins. Mitochondria's role in this process is to be the assassins that ensure the elimination of unnecessary cells between the digits. This is how our fingers are formed. The same process takes place in the brain of an embryo where 50 to 80 percent of the nerve cells that are formed during the early phases of development disappear before birth. Like a Michelangelo beginning with a large piece of marble from which he will carve the sculpture, our body starts with a large mass of cells from which the baby's brain is made. Apoptosis is the chisel that eliminates the excess cells. Through programmed cell death, the life machines help to shape new life.

Mitochondria Fight Viruses, and a Protein Named After a Basketball Team

It's possible that as you are reading this sentence, your mitochondria are quietly helping you neutralize a virus or another invader that found its way into your body. To understand how this works, let's talk about the two subsystems of our immune system—the adaptive immune system and the innate immune system. Many people know that antibodies neutralize specific bacteria and viruses. Antibodies are one part of our adaptive immune system that learns about new invaders and fights them. Essentially, an antibody is a one-trick pony that learns how to recognize and neutralize a specific invader, like a virus or bacteria. The only problem is that learning takes time, and days pass between the invasion and the development of the antibodies. In the meantime, the invading bacteria or virus can proliferate and cause damage.

The innate immune system, on the other hand, acts as our first line of defense. We're born with it—no learning is necessary—so it responds within minutes. It also alerts the adaptive immune system to start working. Our mitochondria play an important role in innate immunity. Part of the innate immune system is a protein with a self-explanatory name: mitochondrial antiviral-signaling protein or MAVS, which is like the burglar alarm at your home. The protein MAVS sits on the mitochondria, ready for action. When some proteins in the cell detect a virus, they bind to MAVS, which signals the cell to make and secrete certain hormones (called cytokines) that activate the innate immune response, and poof . . . the infected cell is dead (along with the virus). This is how our body reacts to the common cold virus, influenza, Ebola, Zika, and many other invaders.

If you're a basketball fan and you wonder if the name MAVS has anything to do with the Dallas Mavericks (known as the Mavs), the answer is that, indeed, it was named by a Mavericks fan. When Zhijian "James" Chen from the University of Texas first discovered this protein, he didn't realize the mitochondrial connection, but when he and his colleagues found that the protein sits on the mitochondrial surface, they looked for a name that starts with an *M*. "At the time, the Dallas Mavericks were playing the Miami Heat in the [2006] NBA Finals," Chen told the *Dallas Morning News*.[5] "That's how we got the inspiration: How about we call it MAVS—Mitochondrial Anti-Viral Signaling?"

The Mavs lost the series against the Miami Heat in 2006 but won against them in 2011. You win some, you lose some, and this is true for the innate immune response, too. While the system works beautifully in many cases, there are also instances where mitochondria can be outsmarted. A recent example is Covid. One of the nastiest characteristics of SARS-CoV-2 (the virus that causes Covid-19) is that it directly attacks the mitochondria-mediated immune response by cutting MAVS into inactive fragments. (Imagine a sophisticated burglar who cuts the alarm system's wires before he breaks into your home.) It is not the first virus to successfully bypass MAVS (another one is the virus that causes hepatitis C), so scientists are studying this vulnerability with the hope of developing ways to fight future sneak attacks. But these are anomalies. Most of the time, the mitochondria are like a good counterterrorism unit that prevents attacks that we'll never know about.

The innate immune system also alarms the body that a particular cell is damaged. When mitochondria's quality deteriorates as we age, or when they experience acute damage from oxidative stress (explained in chapter 3), they leak their DNA into the gooey fluid inside the cell (the cytosol). The cell views this mitochondrial DNA (mtDNA) as a foreign invasion. Mitochondria originated from ancient bacteria and they have their own DNA, and whenever mtDNA leaks into the cell's cytosol, it triggers an innate immune response because of its similarity to the DNA of bacteria.

And sometimes, mitochondria tackle the invader directly. If you have a cat, you probably know that litter boxes should be handled very carefully because cat feces may contain a parasite called *Toxoplasma gondii*. Lena Pernas, a professor at UCLA, showed that mitochondria surround *Toxoplasma* parasites when they invade the cell, depriving them of their food supply. Pernas found that mitochondria increase their fatty acid consumption to starve the parasites of the nutrients they need to replicate.[6]

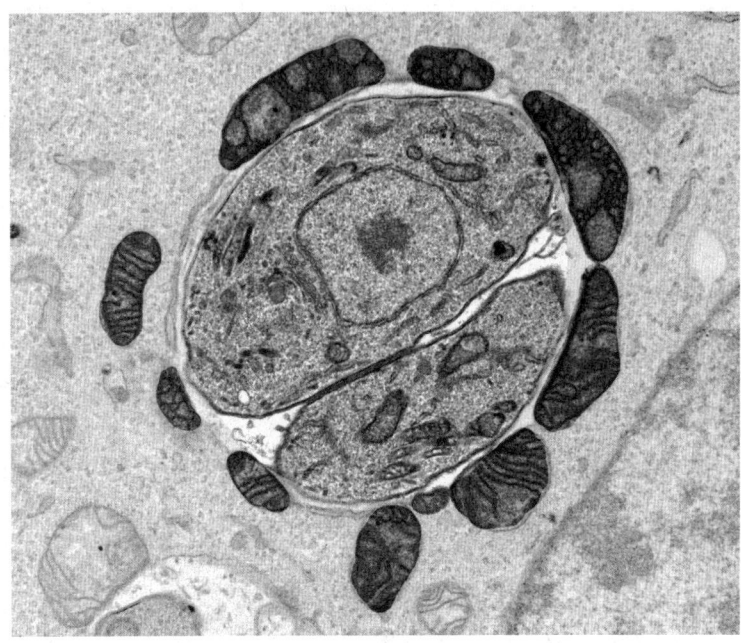

Mitochondria defend us not only from viruses, but also from other invaders. In this electron microscopy image, mitochondria surround parasites called Toxoplasma gondii. For emphasis, the cell's mitochondria surrounding the two parasites are all shaded darker than the rest of the cell. (Courtesy Lena Pernas, UCLA)

Mitochondria as Signaling Stations; And Who's in Charge Anyway?

Mitochondria can also be seen as tiny signaling stations that send and receive messages regarding numerous cell activities. We just saw one example of this in MAVS, the protein that tells the cell to activate the innate immune response. Many signals also come from the Krebs cycle; some of the same building blocks that it produces inform the nucleus (and other cells) about the current state of things. One example is a compound called succinate, which serves as a signal to increase blood pressure by binding to certain receptors in kidney cells. It also further activates our innate immune response.

Mitochondrial signaling is also part of another function that we have discussed—apoptosis, or programmed cell death. How do mitochondria signal to the cell that it's time to die? As we mentioned, in 1996, Xiaodong Wang discovered that mitochondria do so by releasing the protein cytochrome c, which binds with other molecules and starts a signaling cascade telling the cell that it is time to die. Inspired by Wang's findings, a young scientist named Navdeep Chandel became interested in other signals mitochondria may send.[7] One of the first things he looked at was ROS, which stands for reactive oxygen species; it's a by-product of the energy transfer process. You've probably heard about free radicals, which wreak havoc inside the cell; ROS is a class of free radicals that you'll hear much more about throughout the book. In general, ROS can be trouble, and mitochondria have a way to get rid of them, to a certain extent.

Chandel suspected that ROS might also serve as essential signals for different activities. In 1998, his research group showed that a certain type of ROS serves as a signal that has beneficial effects in cases of hypoxia, a state in which the cell is not getting enough oxygen.[8] Chandel and his colleagues found that mitochondria act in this case as oxygen sensors; in the absence of oxygen, they generate ROS, which activate a process to increase oxygen delivery to cells that don't have enough. Subsequently, his group and many others have shown that ROS can serve as signals that activate other similarly helpful processes in the cell.

This brings up an important point about balance in our body and the peril in the "more is better" attitude that some people adopt. As we'll discuss in chapter 7, during the 1990s, when people heard that antioxidants could help our body get rid of oxidants, some started consuming high quantities of antioxidant supplements. Our body—and more specifically mitochondria—produces antioxidants naturally to clean up unnecessary ROS. But if ROS are beneficial in small quantities as signals for certain cell activities, large amounts of antioxidants might *interfere* with this important messaging. More isn't always better.

Mitochondria send ROS as signals to activate another critically important process in the cell called autophagy, which essentially is a recycling program of cell components.

WHAT IS AUTOPHAGY?

To discuss this, here's a little riddle: Every day, an average adult needs to replace 200 to 300 grams of proteins in their body. So how come the daily recommendation of protein is only a fraction of that? The answer is that the cells in your body have an amazing feature called autophagy. Autophagy means "self-eating" in Greek, and indeed, cells constantly chew up damaged bits and pieces and recycle their own components into new cellular structures. And cells do it without interrupting their regular programming. (Just imagine that your car had this feature: recycling old parts into new parts and installing them in place while you're idling at a stoplight.)

Sometimes people get confused between the term we just introduced—*autophagy*—and a term we introduced earlier—*apoptosis* (programmed cell death). To avoid confusion, imagine again that you're the manager of a chemical plant. As we explained, when the machines are beyond repair, you must shut down the plant, and you make sure you do it in an orderly fashion to prevent dangerous interactions between chemicals. This is apoptosis—programmed cell death. However, when the plant is still operating, you walk around the facility on a regular basis to replace pipes or valves that are broken and send them to be recycled. This is autophagy, essentially a maintenance job.

The organelles that execute autophagy are called lysosomes, and studies in the past two decades indicate that mitochondria are key players in regulating this program by sending ROS as signals that activate autophagy. In the simplified view that sees lysosomes as the recycling centers, mitochondria can be seen as the dispatchers that tell the garbage trucks (called autophagosomes) where garbage needs to be collected. If this mechanism is broken, the result is disease. Faulty autophagy can be a feature of some rare mitochondrial diseases, but also of more common ailments like neurodegenerative, cardiovascular, chronic kidney, diabetes, and liver diseases. In contrast, when autophagy is working smoothly, we are healthier. A related term that will pop up throughout the book is *mitophagy*, which is simply the autophagy of mitochondria. This is a fascinating process by which damaged mitochondria are removed, and their parts are "recycled."

A lot of the mitochondrial signals involve communication with the cell's nucleus. This is not too surprising since, as mentioned, a single mitochondrion contains an estimated eleven hundred to fourteen hundred different proteins, and all but thirteen of them are encoded by nuclear DNA. Simply put, a mitochondrion contains many proteins, and the assembly instructions for the vast majority of those are stored in the nuclear DNA. Usually, those proteins are sent to the mitochondria at a steady rate, but when conditions change, the mitochondria send signals to the nucleus with very specific requests. Producing proteins takes energy, too, so there is no point in producing ones that are not needed; our body is very efficient this way. We can imagine a tiny mitochondrion on the phone screaming at the nucleus: "I need just this set of proteins right now. Don't waste my energy on all this other stuff." In more scientific terms, one way that this signal can be sent is through ROS that regulate transcription factors, proteins that turn genes on and off and thus determine which nuclear genes are expressed, at what levels, and at what time. Is it a request or a command?

ROS play an important role in signaling, but they are only one example of mitochondria talking to the nucleus. Throughout the book, we'll see many additional conversations that mitochondria are involved in. They constantly talk with other organelles. They talk with bacteria in our gut. They talk with telomeres, the end caps of our chromosomes that gradually get shorter as we age. Mitochondria are involved so deeply in communication within and beyond the cell that sometimes we start to suspect that they are the ones who are running the show in mysterious ways that scientists are just starting to unravel.

But what does it even mean "to run the show"? In the traditional hierarchical mindset that many of us adopt by default, we tend to think of the brain as managing the body and of the nucleus as managing the cell. But the more we study how our body works, the more we realize that top-down paradigms don't always describe what's really going on. Be that as it may, after reading this chapter, hopefully you agree that mitochondria are far more than just batteries that give us energy. Later in the book, we'll introduce additional tasks they are involved in. It's also very possible that after our book is published, scientists will discover additional mitochondrial functions.

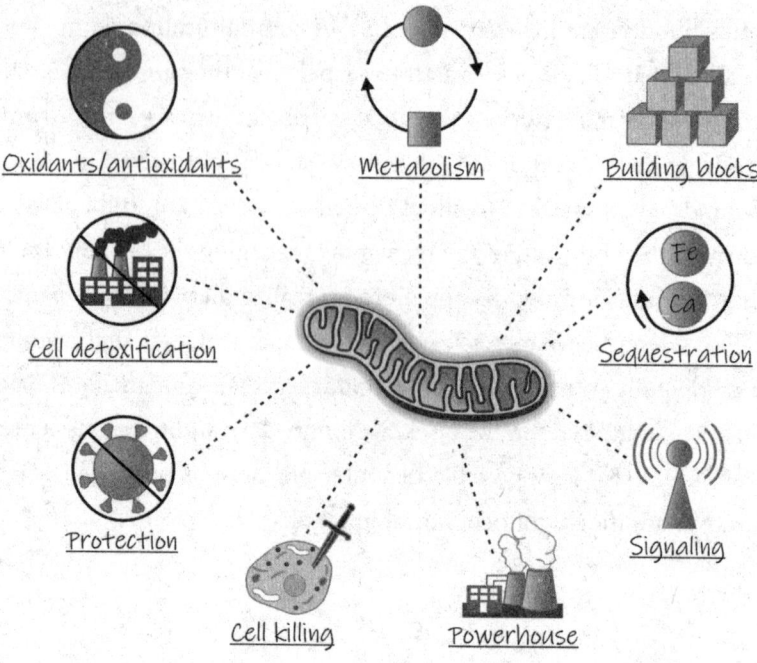

Oxidants/antioxidants   Metabolism   Building blocks

Cell detoxification   Sequestration

Protection   Signaling

Cell killing   Powerhouse

The functions of mitochondria. Mitochondria are mainly known for their important role in transferring energy, which got them the nickname "the powerhouse of the cell." But they do so much more: They metabolize fats and carbs; they provide essential building blocks for our body; they sequester important ions, such as calcium and iron, and release them when they are needed; they send numerous signals; they kill cells that should die; they protect us from viruses and other invaders; they detoxify pollutants that are formed within the body or from the environment; and they keep the fine balance between oxidants and antioxidants.

What Can We Do to Help Our Mitochondria?

Given everything that they do for us, it makes sense to provide mitochondria with the right conditions to thrive. The second part of the book is dedicated to this topic, but here are a couple of broad directions that derive from the concepts described so far. We saw, for example, that mitochondria are metabolic hubs that assemble many of the building blocks of the cell. Clearly, it makes sense to give them the best raw materials for this important job: the right food, at the right amounts, and with the right nutrients. Almost everything that you eat eventually will reach your mitochondria and will either help them or hurt them. If you constantly feed on ultra-processed food, stripped of important nutrients and loaded with sugar, salt, and emulsifiers, you may not give your mitochondria the optimal materials to assemble the building blocks or to perform other tasks we have mentioned. Avoiding toxins in what you drink and in the air you breathe is important, too. We'll see in a later chapter how alcohol, smoking, vaping, and air pollution can disrupt the assembly of some building blocks.

We also saw that mitochondria are signaling stations that report and influence cell activities, and if you want to help them send good signals, exercise is a smart idea. We doubt that we are the first people to tell you that physical activity is good for you, but here is one reason *why* it is so important. There is plenty of evidence that physical activity stimulates mitochondrial signaling. This triggers removal of broken parts and recycling them through autophagy and mitophagy, which help you develop better mitochondria. When you exercise, your skeletal muscles also send beneficial signals to your brain and other organs. This will be discussed at length in chapter 6, and as we point out there, you should consult with your physician before you start an exercise program, and in any case, start slowly but be consistent. Your mitochondria will thank you.

• • •

When having lunch in our backyard, we often marvel at the birds. A sparrow may stand calmly on a branch and take off at amazing speed in a burst of energy. Occasionally, a hummingbird stops by for a drink of nectar from the trumpet vine, and we can't take our eyes off its wings as it hovers by the orange flowers. Talk about energy! The role of mitochondria in providing energy is fascinating, but, again, it would be extremely limiting to see this as the only role mitochondria play in a bird's life. As birds glide over our yard, mitochondria fire signals to regulate their cells' activities and communicate with other cells; they create the building blocks of the birds' bodies; they kill cells that are beyond repair; and they are part of the first line of defense against viruses and other invaders. Mitochondria are wrongly pigeonholed as batteries when in fact they are multitalented multitaskers. This is true for birds, as it is for virtually any creature on this planet, including humans.

CHAPTER 2

On the Move to Help and Heal

The narrow view of a mitochondrion as a power plant brings images of a massive building firmly anchored to the ground. Nothing can be further from the truth; mitochondria are tiny, of course, but more importantly, they are highly dynamic: They constantly change their form, split, merge, move within the cell to wherever they're needed, and even travel from cell to cell. How well they do all this has important implications for our health.

For Daria, the realization that mitochondria move around started with a story of a dead goldfish. In the mid-1980s, as a young research fellow at the University of California, San Francisco, she needed advice on some cell images. People told her to talk to Manfred Schliwa, who was a visiting researcher from Germany. His office was neatly organized except for an empty, dirty fishbowl that Daria was surprised to see on his desk. Noticing the expression on her face, Schliwa's eyes lit up, and he told her that this dirty fishbowl led him to a surprising discovery. Sometime earlier, his goldfish had died, and he almost threw the container away, but being a scientist, he decided to examine the gunk that accumulated inside the fishbowl. Looking at it under the microscope, he noticed some giant amoeba at the bottom, and when he further zoomed in, he noticed some things that were going back and forth on cellular "railways" inside this single-cell organism. What were those moving particles? Reporting his observation in a 1985 article in the *Journal of Cell Biology*, Schliwa speculated that those may have been mitochondria, and studies in the 1990s confirmed that.[1]

In certain cells (like neurons or skin cells), mitochondria move around to wherever energy is needed. For example, if you get a paper cut, skin cells in the center of the cut die, and during the healing process, neighboring skin cells on both edges of the cut will need to move toward each other, something that requires energy. Like fuel supply trucks that move with armored troops that push forward in the battlefield, mitochondria travel to the front within the migrating cells, to supply energy to the part of the cell that is expanding.[2] When your mitochondria are in top shape, this happens nicely, but when they are not, it takes longer. Indeed, as we get older, it takes more time for wounds to heal, which isn't a big deal if it's just a paper cut, but this can lead to significant problems for more serious wounds.

When working correctly, the movement of mitochondria inside the cell is an amazingly coordinated process involving a sophisticated railway system that helps mitochondria get to the right place at the right time. There are no transfer stations in this system and no need to take a cab from the station to the final address. These cellular railroads—known as microtubules—constantly reassemble themselves to connect two points. To move around, mitochondria split into smaller ones and connect to certain protein motors that literally carry them one "foot" in front of the other along these railroads to perform their duty where needed. It's a beautiful system when it's working. But when the system is broken—for instance, in an illness like Parkinson's disease—mitochondria are no longer delivered to the edge of neurons at the right pace, so these edges die and stop communicating with other cells. In some neurons found in the affected part of the brain of Parkinson's disease patients, mitochondria stop moving properly, leading to neuronal death. As a result, the muscle movement of these patients loses its smoothness and is associated with small, shuffling steps (known as Parkinson's gait). Understanding this mechanism fuels new hope for drugs that will recover the broken system. For example, Xinnan Wang, a scientist at Stanford University, has been searching for ways to help mitochondria better latch to the cell's railroads so that they move to where they are needed.

MITOCHONDRIA COME IN DIFFERENT FORMS

This is a good place to break another misconception about mito-chondria: that they are all the same. They are not. While all mito-chondria have many things in common, they don't all look the same, and their behavior can range across organs. For example, while as we just pointed out, mitochondria in neurons move to wherever en-ergy is needed, mitochondria in heart muscle cells don't. Since they are constantly needed to energize the heart's contraction and relax-ation, most of them are tucked in permanent positions across the cell and work their little butts off. There are so many of them that 40 percent of each heart muscle cell is made up of mitochondria.[3]

Another example: While mitochondria in your heart use mainly fatty acids to generate ATP, the mitochondria in your brain use mostly glucose. Neurons can be called for duty at a millisecond's notice, so no time to process fatty acids. This is also why we may faint when our blood sugar is low. Another example showing that not all mitochondria are the same: In sperm cells, mitochondria are coiled tightly around the top part of the tail and generate ATP from glucose to enable the sperm's movement. In contrast, in human eggs (oocytes), mitochondria make necklace–like structures around the nucleus. And even within the same cell, there are different mi-tochondria with different functions: Those around the nucleus may be longer and less active, and those at the periphery of the cells may carry out all the functions that we have described. The bottom line is this: Although all mitochondria have common components, their shape, location within the cell, and function is specific for cell types and organs.

The Dance: They Constantly Split and Merge

There's a photograph on the wall of Daria's office at Stanford that often stumps first-time visitors. It was taken with a powerful electron microscope that can magnify objects more than one hundred thousand times, and it features a figure-eight structure. When Daria asks guests what it is, they rarely identify it as a mitochondrion. They're not alone: When Daria was first shown this photo fifteen years ago and was told this was a mitochondrion, she remembers her succinct reaction: "Huh?" Most of us remember that static picture of a mitochondrion in our biology textbook, but the life machines change their shape all the time. The photograph in Daria's office shows a mitochondrion going through fission, the process by which more mitochondria are generated. And mitochondria also merge with other mitochondria, a process known as fusion. Together, through this constant dance of fission and fusion—splitting and merging—the mitochondria's ability to function properly is maintained.

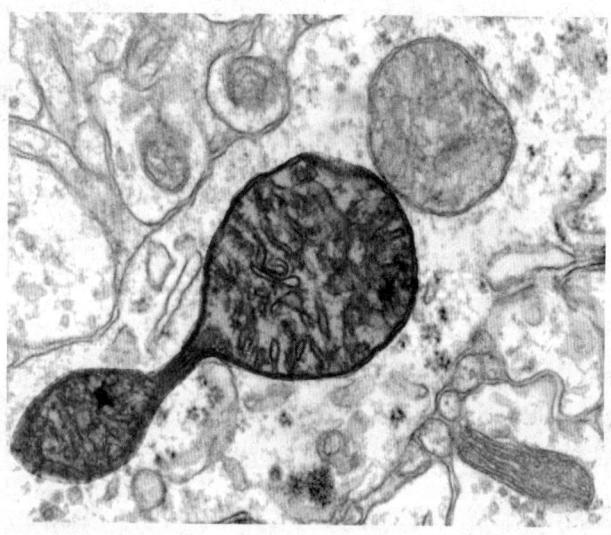

A mitochondrion going through fission. In this electron micros-copy image from a rat's brain, a mitochondrion is seen splitting into two mitochondria. Fission may precede the process of mitophagy through which damaged parts of mitochondria are removed, which keeps the remaining mitochondria healthy. A lot of fission and mitophagy happens at night, and when you get enough quality sleep, you give your mitochondria a chance to complete this cleanup process. (This is the image that decorates Daria's office. For emphasis, the mitochondrion going through fission is shaded darker than the rest of the image.)

The dance of fission and fusion is a fascinating process. Consider, for example, how a single mitochondrion decides whether it should merge with another mitochondrion. It starts with a handshake, during which each mitochondrion seems to check out the other one, as in speed dating. If either of them senses that the other one is not good enough, they move on, looking for a better partner.[4] On the other hand, if merging is beneficial, they connect, fuse, and start to exchange components to complement each other. Think about how this ingenious process helps the state of mitochondria in the cell and thus your health: A mitochondrion that is repeatedly rejected by others doesn't get the good stuff from other mitochondria and eventually will be removed through mitophagy, the recycling program of mitochondria. In contrast, good mitochondria get better.

The process of fission, the division of a single mitochondrion into two, is no less fascinating. Fission is the way more mitochondria are generated, which is needed, of course, when the cell itself divides, which happens for some cells every twenty-four hours. But even when the cell does not divide, mitochondria need to split to remove damaged parts, in the same way that a tree needs to be pruned for optimal growth. This happens through the mitochondria's recycling program—mitophagy. Remarkably, before mitophagy can begin, our body separates between what needs to be recycled and what can still be used. Recall that mitochondria have their own small DNA (mtDNA). Unlike nuclear DNA, which has efficient machinery to prevent and correct mistakes, the mtDNA is much less efficient in taking care of such errors. This means that the mtDNA collects many mistakes throughout the life of each cell. But here's the good news: Unlike nuclear DNA, each mitochondrion has several copies of mtDNA. Therefore, in the same mitochondrion, you may have some lucky mtDNA that are still in good shape, and less fortunate ones that were damaged. The same is true for the proteins in a mitochondrion. How is it then that mainly damaged mtDNA and damaged proteins are removed from a mitochondrion? This is yet another extraordinary process that happens inside our mitochondria. Damaged components assemble at one end of a mitochondrion while the intact mtDNA and proteins stay on the other end. Next, in coordination with another organelle called endoplasmic reticulum (ER), something like a noose is formed around the mitochondrion and cuts the mitochondrion into two mitochondria. The part that was pinched off contains the bad stuff while the other part is a healthy fresh mitochondrion ready for action. The bad section is collected by autophagosomes, which we can think of as garbage trucks, and brought to lysosomes, which we can think of as recycling centers. There, the damaged parts are digested into their building blocks, and those are later used to build new cell components.

How is this fascinating process relevant to your health? Here's an example: During the day, mitochondria focus mainly on producing ATP and building blocks for all your daily activities; at night, while still generating ATP, they switch to a recovery mode—sorting the functional parts from the damaged ones and removing the damaged parts through mitophagy. So when you get enough quality sleep at night, you give your mitochondria a chance to complete this pruning process. Part of the "noose" that cleaves a mitochondrion into two mitochondria is a protein called Drp1, and researchers found that it receives signals from (and sends signals to) our body's circadian clock (the internal clock that tells us when it's time to sleep and when it's time to get up).[5] We'll go into more details in chapter 9, but the key point is that getting enough quality sleep allows mitochondria to go through this deep cleanup. You should get enough sleep at night to let your mitochondria complete the job.

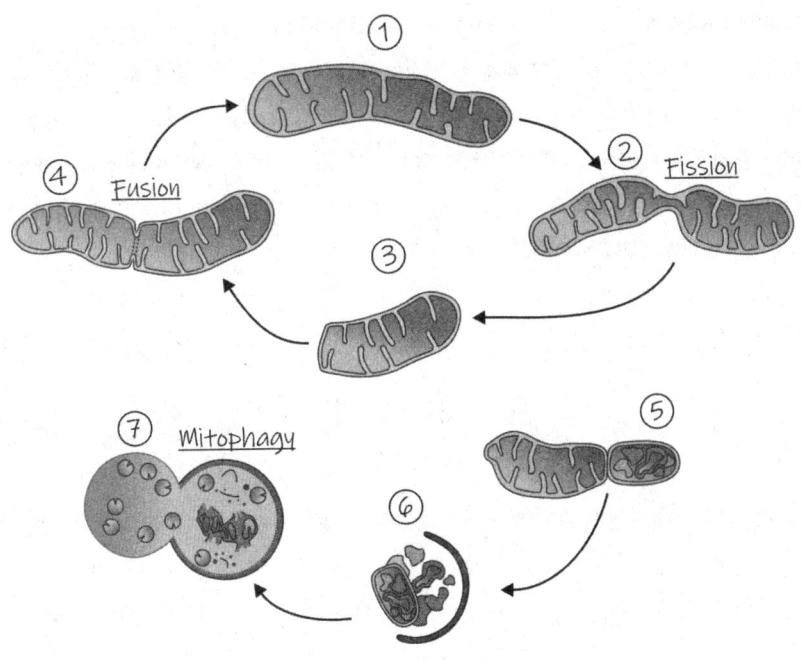

Mitochondria are dynamic. As part of their dynamic nature, mito-chondria go through a process of fusion and fission. For example, a mitochondrion **(1)** can go through symmetric fission as can be seen in step **(2)**. The new mitochondrion **(3)** performs its job in the cell, and at some point, it may meet another mitochondrion and fuse with it **(4)** to exchange components. In some cases, a mitochondrion can go through asymmetric fission to prune itself and get rid of dam-aged mtDNA and proteins **(5)**. After this happens, one part is a healthy fresh mitochondrion ready for action. The part that was pinched off is collected by autophagosomes **(6),** which we can think of as garbage trucks. The pinched part is then brought to lysosomes **(7)**, which we can think of as recycling centers, where it goes through mitophagy.

We opened the chapter by highlighting the importance of flawless mi-tochondrial movement in maintaining our health. We saw, for example, that poor mitochondrial movement within the cell can lead to slow wound healing.

This same idea can be seen with fusion and fission. These processes are essential to life, but too much fission isn't good because the mitochondria become too small to function properly and instead cause further cell injury. Excessive mitochondrial fission was reported in neurons of patients with neurodegenerative diseases, and this took Daria and her team on a drug development journey that we will discuss in chapter 5.

HISTORICAL PERSPECTIVE

Before we move on to discussing another facet of the dynamic nature of the life machines, here is some historical perspective. While some details that we outlined here are cutting-edge science, we don't want to create the impression that early researchers thought about mitochondria as a static organelle. The very term mitochondrion consists of two words in Greek: *mitos* (thread) and *chondros* (grain), which suggests that Carl Benda, the German scientist who coined the term in 1898, understood that mitochondria can have different shapes. Sometimes they look like spaghetti, and sometimes like tiny pieces of ground beef in Bolognese sauce. There are also drawings from the mid-twentieth century depicting different shapes of the organelle,[6] but it seems that the cartoon of the bean-shaped organelle with the wiggly line in the middle is stuck in people's minds. We hope we have helped to change this misconception.

They Move (or Send Stuff) to Other Cells

Here's a surprising fact: The state of the mitochondria in your skeletal muscles may be a good predictor of cognitive impairment. In 2023, researchers from the National Institutes of Health found that the condition of mitochondria in our legs is associated with biomarkers (biological signs) of Alzheimer's disease and neurodegeneration.[7] They analyzed data of 469 individuals who participated in the Baltimore Longitudinal Study of Aging, which (among other things) measured mitochondrial functions in thigh muscles. The study also included cognition tests, and when the researchers reviewed the results collected over approximately five years, they found that lower mitochondrial function in the muscles was associated with an increased risk of developing mild cognitive impairment or dementia. In contrast, study participants whose muscle mitochondria were functioning well were less likely to develop cognitive impairment. How can this link be explained? As of this writing, mitochondria's involvement in the link between the brain and skeletal muscle is not fully understood, but here are a couple of possible pieces of the puzzle. First, as outlined earlier, mitochondria send signals to other cells, impacting other tissues and organs across the body. Second, recent studies show that in addition to sending signals, mitochondria can get out of one cell and move into another!

Do mitochondria travel from your thighs to your brain? We are not aware of a study that demonstrates this, and the link remains a mystery. But something's going on, and this and similar findings suggest that we should think more broadly about mitochondria's influence throughout the body. Mitochondria have been seen for decades as homebodies with their influence limited to a single cell. But more and more studies show that mitochondria can move out of their original cell or send compounds that influence the functions of other cells in remote organs. Mitochondria that leave a healthy cell support the life of the cells that they enter. Damaged mitochondria that leave stressed cells can alert neighboring healthy cells about danger. If taken up by the healthy cells, these damaged mitochondria can be degraded but may also overwhelm these recipient cells. In other words, high-quality mitochondria can help other cells, while low-quality mitochondria may kill them.[8] You may ask yourself: What do you mean by "high-quality mitochondria" and where do I get those? High-quality mitochondria are ones that, for example, are efficient at generating ATP and building blocks, and that produce enough antioxidants to get rid of toxins. You can't buy them at a drugstore, but you can improve the condition of your mitochondria through lifestyle choices that we will describe in part II.

The main takeaway, again, is that we need to think more broadly about the impact of mitochondria on our health. Since they can move out of their original cell or send compounds that influence other organs, we must realize that the benefits of having healthy mitochondria are not limited to a single cell. Here are just three more examples of how mitochondria help other cells.

The first example occurs in our brain. In 2014, scientists at the Johns Hopkins University School of Medicine observed that neurons send damaged mitochondria to astrocytes, which are support cells that help neurons in our brain.[9] Two years later, scientists at Harvard Medical School reported that mitochondria also go in the other direction—from astrocytes to neurons. Experiments in mice suggested that during a stroke, astrocytes help damaged neurons by sending them fresh mitochondria.[10] Simply put, it appears that when these cells notice that their neighbors are in trouble, they inject them with some life machines that provide them with ATP, building blocks, and all the other good stuff that we described in the previous chapter. (As we'll see in chapter 5, this may open new therapeutic opportunities, and some scientists and physicians try to emulate these natural processes to treat the heart, brain, and other organs.)

The second example is about what happens in our body when we bleed. In each drop of blood there are about 3 million platelets, and each of these platelets carries five to eight mitochondria that they dump into the critical area, leading to millions of extracellular mitochondria at the site of the wound.[11] Why are mitochondria needed here? First, to ensure healing. During injury, blood vessels are damaged, and the body immediately starts to develop new blood vessels to replace the severed ones. The extracellular mitochondria provide some of the building blocks for this effort. Second, mitochondria provide extra energy to certain white blood cells called neutrophils that start the fight against bacteria that may get in through the wound.

The third example illustrates that cells donate mitochondria to other cells when we're healthy, too. For example, macrophages are important immune cells that work around the clock to protect us. All this hard work damages their mitochondria, and it turns out that fat cells send them fresh mitochondria so that those macrophages don't get exhausted.[12]

When Mitochondria Send Little Treats

To help a friend, you sometimes drive to visit them at home, and we just saw that mitochondria do this by traveling to other cells. But sometimes it's easier just to send a care package, and it turns out that our mitochondria do this, too, by secreting certain peptides, which are tiny fragments of proteins. It's as if they send "little treats" to help other cells and thus improve our health and possibly extend our health span (the portion of our life when we are in good health, not just alive).

The secretion of these peptides depends on your genetics, but also on your lifestyle. Consider, for example, one such peptide called humanin. Research has demonstrated (in animals) that the peptide can play a protective role against various diseases, including neurodegenerative diseases, diabetes, and cardiovascular diseases, which suggests that having humanin in your body is a good thing.[13] One study showed that the blood levels of humanin in adult children of centenarians were three times higher than the level in the control group consisting of people the same age but whose parents died before they reached one hundred. This suggests that some lucky people may be genetically disposed to have more humanin than the rest of us, which means that they may live longer. And here's the potentially good news that we'll further discuss in chapter 6: Recent studies suggest that certain types of exercise moderately increase the level of humanin in our body.[14]

Here's another example for a peptide that derives from our mitochondria, and it, too, has been reported to be secreted in humans during exercise. The peptide is called MOTS-c, and it has several potential benefits, including reducing obesity and extending our health span. Researchers at the University of Southern California showed that treating aged mice with MOTS-c increased their physical capacity and health span.[15] These findings may be related to health span in humans as well: Japanese supercentenarians (individuals who are 110 or older) express a unique type of MOTS-c that may explain the unusual longevity of these people.[16]

One takeaway on a personal level is obvious: Exercising may stimulate our mitochondria to send beneficial peptides throughout our body. But there is another important takeaway. Scientists believe that it can lead to the development of new drugs to help humans. For example, Pinchas "Hassy" Cohen from the University of Southern California, a pioneer in the field, believes that mitochondrial-derived peptides can lead to the development of therapies for age-related diseases such as type 2 diabetes and Alzheimer's. As you remember, mitochondria have their own DNA (mtDNA), which we'll discuss further in the next chapter. And researchers have identified mutations in mtDNA-encoded peptides that are associated with certain diseases. For example, a mutation in MOTS-c is associated with a higher risk of diabetes; a mutation in humanin is associated with a higher risk of cognitive decline; and a mutation in a peptide called SHMOOSE is associated with up to a 50 percent higher risk for Alzheimer's disease.[17] It is possible that the deficiency driven by such mutations can be treated by providing active peptides or compounds that mimic their effects.

For decades, the common belief among scientists was that the mitochondrial DNA (mtDNA) encoded only very few proteins, and everything else that was produced under instructions from mtDNA was junk. The discoveries in this field in the past two decades have demonstrated that those peptides are not junk at all.[18] We have a feeling that we have not heard the last word on these "little treats."

What Does This Mean to You and to the Future of Medicine?

Mitochondria dancing, moving within the cell, out of the cell, and sending signals throughout the body. All these discoveries may have profound implications on the way we think about medicine and wellness. Medicine is currently organized around anatomy, with specialists for every organ. We see a cardiologist for our hearts, a gastroenterologist for our gut, a dermatologist for our skin, a nephrologist for our kidneys, and so on. But if mitochondria are moving around, if they send signals to other organs, if they function as the conduit between organs, then perhaps we should be thinking about medicine differently and paying much more attention to these tiny organelles. As we'll see in chapter 5, taking care of mitochondria through medical interventions can affect multiple organs in our body.

It also means that by caring for your mitochondria through small lifestyle changes, you are caring for your entire body. We touched upon just two examples of that in this chapter: first, exercise triggers mitochondria to send little treats that affect multiple organs; second, a good night's sleep allows mitochondria in your brain to go through deep cleanup. We will delve into these and additional practical applications in part II, but the main takeaway for now is this: The effect of taking care of your mitochondria is not limited to one cell or one organ; it has widespread effects on your body.

By now, we hope you appreciate the importance of mitochondria, their dynamic nature, and the many roles they play in our body. Over the years, however, their functions naturally deteriorate, so our mitochondria need all the help we can give them. We believe that understanding this is the key to healthy aging, which we discuss next.

CHAPTER 3

The Key to Healthy Aging

In 2017, after a short sabbatical in Kyoto, we decided to go for a five-day hike along the Kumano Kodo, a trail that has been used for over a thousand years by Japanese pilgrims. The first morning was magical. Tall Japanese cedar trees on both sides of the dirt path marked our way, as we passed through tiny villages and near small shrines. Ferns sprouting next to rotted wood reminded us that one of the goals of this pilgrimage was to be ritually reborn and rejuvenated. Our plan for the first day was to walk to the Hongu Taisha shrine, visit the site, and have lunch there. From that spot, it would be a short hop to Yunomine Onsen, a village famous for its hot springs, where we'd spend the night. The first leg of our hike worked according to our plan, and we reached Hongu Taisha earlier than expected. What came next was a reminder that humans are not that great at seeing what's around the corner.

Fooled by our good progress, we took our time at Hongu Taisha, browsing gift counters, playing with pebbles at the nearby river, and sitting down to eat udon noodles at a small restaurant. After the late lunch, we agreed that it was time to move on, and after another hour of wasting time looking for the beginning of the trail, we started walking toward Yunomine Onsen. It was a steep climb. In contrast to the morning's flat and clear path, this winding trail was scattered with obstacles: slippery areas, small boulders, and our least favorite—thick exposed roots one can easily trip over. Then, shortly after sunset, the forest went dark, as though someone had suddenly turned off the lights. We were now climbing a steep mountain, with no flashlights, just the light of our iPhones, with a pretty good chance we'd end up with a sprained ankle or worse. There was no one around. "Slowly," we kept reminding each other, sometimes in unison, while looking for a safe spot to place our hiking poles. A day that started as a stroll along a clear path ended with us inching our way uphill in the dark in a remote Japanese forest. The final stretch wasn't what we expected.

We've been married for more than forty years, and we try to do all the right things: We exercise, we walk, we eat healthy food, we sleep well. And we often wonder: How much time do we have left? And more importantly, how will we spend the last decades of our lives? Will we be able to maintain our good health, or will we spend more time in bed and doctors' offices, frail and weak? What is our final stretch going to be like? Like most people, we want to have a long healthy life, or to put it more academically, we don't just want to extend our lifespan, but also our *health span*.

Mitochondria as We Age

In 2016, two women entered the lobby of a research building in Leiden, a city in the Netherlands. They were both around seventy years old, had the same height and build, and were dressed nicely. One woman, we'll call her Emma, walked slowly with a little hunch, let the receptionist know she was there, and sat down in the waiting area. The other woman, we'll call her Sophie, walked much faster, checked in, and leaned against the wall to do some stretching. Emma and Sophie were two of twenty elderly men and women who volunteered to participate in a study to assess the health of their mitochondria. Half of these persons were active like Sophie, and the other half were medically defined as "pre-frail" like Emma (characterized by things like slow walking speed, low muscle strength, or sedentary behavior).

Several tests were performed on these volunteers, including a biopsy of their thigh muscle, and the results showed a clear difference: The mitochondria of active people like Sophie were in much better shape than the mitochondria of "pre-frail" folks like Emma. The researchers found "a striking association of the development of pre-frailty with a decline in skeletal muscle mitochondrial function."[1] A study from Johns Hopkins University showed similar results in two small groups of eighty-year-olds. Half of them were frail and half non-frail, and the mitochondria in the skeletal muscles of the latter group were functioning much better.[2] Bottom line: Having mitochondria that are in good shape is associated with being healthy later in life.

To be clear, the difference between the two groups, or between Emma and Sophie, may be attributed to their lifestyles—there's plenty of evidence that physically active people have better mitochondria—but it can also be attributed to their genetic makeup. Researchers point to genetic differences between frail and non-frail individuals. It's not that there is a "frailty gene," but for example, someone who's genetically prone to low bone density (which can lead to osteoporosis), is more likely to break a bone, a path that can lead to frailty. So it's clear that it isn't only about what we do, but also about the cards we've been dealt.

Genetic advantage or not, we don't fool ourselves: Even people who live the healthiest lifestyle must face reality, and the reality is that mitochondrial functions deteriorate as we age. Yet it is also true that healthier older people tend to have healthier mitochondria. So when we think of our own aging, we are determined to do everything in our power to maintain the health of our mitochondria. We believe that this is good advice for anyone, regardless of their age, and it certainly doesn't apply only to people around seventy like Emma and Sophie. In fact, a study of 146 healthy men and women aged eighteen to eighty-nine showed that different measures of mitochondrial functions (including ATP production) declined linearly with age.[3] In other words, thinking about your mitochondria shouldn't begin when you qualify for a senior discount. The earlier you start, the better.

There are good reasons to believe that taking care of your mitochondria in the early stages of your life will help you in your final stretch, and we think of mitochondria as the key to healthy aging. Staying physically active, eating healthy food, getting enough sleep, and reducing stress are some of the ways to be good to your mitochondria, which we'll discuss in the second part of the book. That evening on the Kumano Kodo trail, we were thankful for all the squats, lunges, and leg curls we'd done over the years to strengthen our legs, and especially for the one-leg stands and other balancing exercises. Those became helpful when we reached the top of the mountain. We'll never forget the descent toward Yunomine Onsen, hearing the voices of people in the distance, and the great sense of relief when we were finally there. With steam rising from the hot river that runs through the village, and the pungent odor of the hydrogen sulfide in the water, the place reminded us of depictions of hell, but we were in heaven. The following day, we bathed in an ancient hot spring bath rich in hydrogen sulfide; it was an unforgettable day. In the end, this is what we ask for: to be able to have such experiences for as long as we can and shorten the period when we'll be confined to our home or, worse, to our bed. We know that the last stretch will have its challenges; the obstacles will be there. We also know that we can help our mitochondria help us better deal with those hurdles.

An oversimplistic view of the link between mitochondria and aging goes something like this: Mitochondria provide energy, and because they deteriorate over the years, old people have less energy. While this is not wrong, there are two problems with this explanation. First, the expression "have no energy" is often understood as being tired or being lethargic, but all the organs in our body run on ATP, and a decline in ATP can be manifested in a wide variety of ways that are much more serious than just being tired. The second problem with this oversimplistic view is that mitochondria do much more than energy transfer, so mitochondrial dysfunction isn't just about having less energy.

The fact is that after the age of sixty, many people become ill with one of the following diseases: cancer, diabetes, cardiovascular, or neurodegenerative diseases. These are the ailments that end up killing most of us. A growing body of evidence associates these diseases with the quality of the mitochondria, so in addition to the apparent changes in energy, hair color, skin tone, and eyesight that occur as we age, the most concerning change is that we are more likely to get one of those major diseases. We'll talk more about this in the next chapter, and as we'll explain, there are still many open questions about the association of each of these diseases with mitochondria. For now, our point is that healthy mitochondria can delay the onset of these ailments and decrease their severity. By tending to our mitochondria, we reduce the burden on the cell and give it a better chance to heal and help other cells.

Studies show that virtually all the functions of mitochondria that we listed in chapter 1 can deteriorate over time: the production of building blocks, the killing of irreversibly damaged cells, the ability to fight viruses, the ability to send numerous signals, and of course, ATP production. It's not that these functions stop working one day, but they do slow down as we age. In chapter 1 we also briefly mentioned that mitochondria talk to telomeres, the end caps of our chromosomes, which gradually shorten as we grow older (like the plastic pieces that prevent the unraveling of our shoelaces, telomeres keep the chromosomes' ends structurally intact). Now, scientists know that damaged mitochondria shorten the telomeres, and conversely, critically short telomeres damage our mitochondria.[4] There are also changes in the mitochondria's dynamic behavior, the balance between growing longer via the fusion of two mitochondria or becoming shorter via fission into smaller mitochondria (discussed in chapter 2). When this fission/fusion dance becomes less graceful, the balance suffers, damaged mitochondria accumulate, and these wreak further havoc. Additionally, as we age mitochondria don't move as quickly within the cell, which is associated with a variety of problems ranging from slower wound healing to Parkinson's disease. Mitochondria in the elderly also become less efficient in helping other cells by sending the "little treats" we've discussed earlier.

All the above describes *what* happens as we age, but the really important question is *why*. Why do mitochondria deteriorate in our older years? We discuss this in detail next, but on a conceptual level, we can say that it's a matter of errors and pollution. Think of a factory. In its early days, the machines were in pristine condition, running smoothly. They created pollution, but any buildups of soot or rust were removed right away by the crew, and an excellent management team was always on top of things. However, after years of operation, things change. The cleanup of soot and rust is not as efficient, and the buildup starts to accumulate and slow down the machines. In addition, the once brilliant management team makes more and more mistakes. Things start to break down.

This metaphor obviously oversimplifies things, and in the next few pages, we'll describe the errors and the pollution that happen in mitochondria in our later years, errors in the mitochondrial DNA and pollution from reactive oxygen species (ROS).

What happens to mitochondria as we age? When we are young, mitochondria function like new factories run by brilliant and efficient managers (the mitochondrial DNA). As we get older, like in old factories, things start to break down and the "managers" make more and more mistakes (increased errors in the mitochondrial DNA). So as we age, the cleanup of "soot and rust" (ROS and other toxic molecules in mitochondria) and replacement of broken parts (by mitophagy) are not as efficient, slowing down the machines, causing decreased ATP and building block production and increased inflammation.

The Mitochondrial DNA and the Safe Zone

First, let's talk about the "management team," the mitochondria DNA. When most people hear the term *DNA*, they think of nuclear DNA, the main instruction manual for building your body and maintaining it. This is understandable because of its central role in the cell. But like the cell nucleus, mitochondria have their own DNA (mtDNA), which is much smaller and circular. Mitochondria are the only organelles (other than the nucleus) that have their own DNA, and to understand the origin of this unique trait we need to go back to the 1960s, to a woman who reshaped our understanding of cellular evolution, Lynn Margulis.

In 1967, Margulis wrote a paper that was so revolutionary that it was rejected by fifteen different journals before it was finally published.[5] In this paper and subsequent publications, Margulis pointed out some similarities between mitochondria and certain microbes, proposing that about 1.5 billion years ago, simple bacteria merged with some early cells and formed a mutually beneficial relationship with their hosts. Those bacteria knew how to transfer energy with the help of oxygen, and the host cells provided protection. Over time, Margulis argued, these bacteria became what we now know as mitochondria. The idea was initially rejected because it went against prevailing beliefs at the time, but with the emergence of new research tools in the 1970s and 1980s, other researchers started to confirm her hypothesis by conducting their own research. Today, Margulis's theory on this issue has been accepted by most scientists in the field. This theory has profound implications. It means that without mitochondria, life as we know it on planet Earth would not have existed. No plants, no animals, no humans. The life machines enabled life.[6]

However, the fact that mitochondria originated from bacteria does not mean that these organelles can leave our body and live independently. Mitochondria can no longer live without us, and we cannot live without them. As mentioned earlier, a single mitochondrion contains more than a thousand proteins, and most of them are encoded by nuclear DNA. Simply put, over millions of years, mitochondria "delegated" to the nucleus most of its genome. But (and this is a big but), mitochondria kept control over thirteen proteins and the machinery to express them autonomically. The thirteen proteins are a sort of linchpin for the functioning of the organelle. This means that when mtDNA is damaged, it doesn't produce critical mitochondrial components properly. Why is damage more likely as we age? The mtDNA keeps replicating, and every time it replicates, there is a chance that an error in the sequence will be introduced. The machinery that ensures fidelity of the DNA during replication in the mitochondria is not as good as that in the nucleus and therefore errors in the mtDNA accumulate over time. To use our factory metaphor, the management team repeatedly messes things up.

When you were born, the DNA copies in your mitochondria were in close to pristine condition. (There are exceptions that we'll discuss later.) But over the years, this has changed, and you accumulated damage in some copies of the mtDNA, because the machinery that sorts damaged from pristine is overwhelmed. The good news is that we can live with some damaged copies of mtDNA and be fine. The trouble starts when the percentage of damaged mtDNA in a cell goes beyond a certain threshold. It is then that disease manifests itself, and this is a concept that is critical in both aging and disease.

For example, a woman who carries mitochondria with a serious mutation, such as a large deletion in the mtDNA, may appear healthy and her children are not necessarily doomed to be sick. Each cell can have hundreds or even thousands of mitochondria, and each mitochondrion has multiple copies of its DNA. If the number of mitochondria with mutated DNA is under a certain threshold (generally considered to be 60 to 80 percent of the total copies of mtDNA), the disease will not manifest. So a mother who carries a mitochondrial mutation can have one kid who has the disease and another kid who doesn't have it (because the number of mutated mitochondrial DNA that happened to be packaged in the egg that produced the healthy kid did not pass the threshold). Moreover, because of the random distribution of the mitochondria to different cells as each organ is formed within the same kid, a primary mitochondrial disease can present itself in one part of the body at an early age, and show up in another part years later as the number of mitochondria with mutated DNA in that new part of the body passes the threshold.

Is this true for age-related diseases as well? In other words, do certain diseases manifest themselves only once "bad" mitochondria cross a certain threshold? This has been established in primary mitochondrial diseases, but the same concept is likely at work with age-related diseases. For example, researchers showed that mtDNA mutations that accumulate as we age can cause a decrease in ATP production and an increase in cell dysfunction.[7] Further support for the relevance of this concept in common diseases comes from a cutting-edge method that we'll discuss in chapter 5—mitochondrial transplantation. In this procedure, healthy mitochondria are injected into a patient's organ or blood vessel, and this method shows promising results in a range of diseases from cancer to neurodegenerative diseases.[8] Even though the number of pristine mitochondria that are injected is very low (relative to the billions of mitochondria in the body) results show that it is often sufficient to provide benefit. A likely explanation for the positive results is that the extra number of healthy mitochondria pushes the number of bad ones under the threshold.

The phenomenon described above, in which a single cell has pristine as well as damaged mtDNA, is known as mtDNA heteroplasmy, and it highlights a key point when it comes to therapeutics: We don't have to cure every single mitochondrion to improve our health. We just need to bring the number of damaged mitochondria under the threshold. It's also a fundamental point in thinking about prevention: We can live with some imperfections in our mitochondria, but we want to keep them under the threshold. mtDNA heteroplasmy isn't a lab test that your physician can prescribe, at least not yet, but it can provide a useful framework in thinking about staying healthy as you age. In fact, one can look at the threshold concept more broadly, and during our long walks in writing this book, we started talking about staying healthy as paddling a canoe in a large lake.

Imagine that you are on vacation canoeing on a beautiful lake, and you paddle gently while your eyes move from the deep blue water to the lush green vegetation surrounding the lake. The water is calm, with a very slight current toward the south bank of the lake. When you rented the canoe, the guy at the rental booth told you to make sure the current doesn't take you too close to the south side of the lake, where there's a dangerous waterfall. You want to stay away from that area; if you get too close, you may find yourself in rapid waters that lead to the waterfall. If you get to the rapids, it may be too late. You want to stay in the safe zone. The safe zone of the lake represents your healthy state, and the waterfall represents a pathological condition that can happen on a small or larger scale. Paddling represents actions you take that may help you stay in the safe zone or, conversely, may get you closer to the dangerous waterfall. Examples of paddling that will keep you in the safe zone include exercise, a good night's sleep, and a healthy meal. In contrast, habits like smoking, vaping, eating fried food, and drinking alcohol will get you closer to the waterfall. You don't need to paddle frantically, but you want to consistently paddle away from the zone of no return.

The link between mtDNA and disease is another concept that wasn't easily accepted in the scientific community at first. "When I started this work in 1971, it was definitely thought to be a completely foolish endeavor," Doug Wallace, a pioneer in the field, told us. As a young scientist, Wallace wanted to study how mitochondrial DNA may relate to disorders. In 1963, Margit Nass and her husband, Sylvan, at Stockholm University had seen DNA inside mitochondria using electron microscopy,[9] and Wallace wondered if this DNA could mutate. "Nobody thought that was a good idea," Wallace said, except for Joshua Lederberg, the chair of the Department of Genetics at Stanford, who was willing to gamble that mitochondrial DNA might be interesting. He hired Wallace, who started recruiting families from the Stanford community and analyzing their mitochondrial DNA. When Wallace looked at the results, he thought he had made a mistake because he was expecting it to segregate according to Mendel's heredity theory: half from the father and half from the mother, but this did not happen. Wallace repeated the experiment in several different ways before he was confident: mitochondrial DNA was inherited only from the mother.[10]

To use us as an example, this means that our children have Daria's mitochondrial DNA, which she inherited from her mother, Alda, who inherited it from her mother, Cesarina, who inherited it from Virginia, who got it from Gemma, who inherited it from Elvira Barbieri, who lived in Siena, Italy, two hundred years ago. What about Emanuel's mtDNA? For his part, Emanuel carries the mtDNA he inherited from his mother, Mirjam, who got it from her mother, Lucie, who got it from Nettchen, who got it from Ricca, who inherited it from Rosa, who lived in Brilon, Germany, two hundred years ago. When our children were conceived, Emanuel's sperm was loaded with mitochondria that fueled the tight race to Daria's egg. But even though they worked their tails off, the mitochondria in his sperm were destroyed once Daria's egg was fertilized.

Of course, the mtDNA that we carry goes much further than Italy or Germany. Daria still remembers how in the mid-1980s, at a retreat of the Department of Biochemistry at UC Berkeley, a young post doc named Rebecca Cann presented a revolutionary idea that built on the work of Doug Wallace and others[11]: Cann and her mentor, Allan Wilson, argued that we all inherited our mitochondrial DNA from a single woman who had lived in Africa two hundred thousand years ago. When Cann and Wilson published their paper in 1987, the media went wild with the idea of "mitochondrial Eve."[12] Today we know that there were probably several such Eves, and we can trace our own mtDNA to Africa through some genetic-testing services.

Beyond the curiosity about our ancient origins, mtDNA is relevant to aging and disease. By sequencing the mitochondrial DNA of indigenous populations, Wallace and other researchers in the field found that people from different geographic origins carry mitochondrial DNA common to that region. Such an inherited mtDNA sequence unique to a specific ancestry group is called mtDNA haplotype, and a group of related haplotypes is called a haplogroup. Wallace believes that different haplogroups made our ancestors better suited to the corresponding local environment. For example, a woman in Africa had to be able to run away from lions, so she needed the most efficient mitochondria. In contrast, women who migrated to northern Europe had to defend themselves from the frigid environment, so through natural selection, mitochondria that produce relatively more heat became the common haplotype. This may explain why Kenyan marathon runners frequently win races, and Wallace jokes that it may also be why he and his wife argue about the temperature in their bedroom; Wallace's wife is of Scandinavian origin, and he is not. She's always hot, and he's always cold.

Such different mitochondrial DNA lineages also explain things related to aging and disease. For example, by analyzing the mtDNA of 112 Japanese over the age of 105, researchers identified a particular haplogroup that is a marker for extreme longevity in Japan.[13] It also explains why certain ethnic groups are more susceptible to certain diseases. Certain mitochondrial DNA lineages are strongly related to diabetes, blindness, deafness, and renal failure. For example, there is a mitochondrial lineage that is associated with diabetes in Asians and Asian Americans.

Going back to the topic of aging and to our factory metaphor: So far, we discussed the management team of the factory that stumbles again and again as we age. This represents damage to mtDNA, which happens over time through replication errors that lead to mutations, and the instruction manual that the management team received to begin with (haplotype). Now let's move on to the other problem that occurs in this factory—pollution.

The Awe-Inspiring Engine (and the Pollution It Creates)

The second source of trouble that hurts your mitochondria is cellular pollution, most commonly in the form of reactive oxygen species (ROS). As we saw in chapter 1, ROS are free radicals that are created mostly inside the mitochondria.

Although ROS are sometimes beneficial (passing critical information from one cell to another), they usually mean trouble and can wreak havoc in the cell.

How are ROS created? To discuss this mitochondrial-generated pollution, we should go more deeply into how mitochondria produce ATP and introduce one of the most amazing machines in the human body. We're talking about those tiny rotors we mentioned earlier that are spinning inside your mitochondria. If you could magically shrink and visit a mitochondrion to see those rotors in action, you'd be able to observe them spinning at a dizzying speed, churning out molecules of ATP. These are actual rotors. The scientific name for these rotors is ATP synthase (because the rotors synthesize ATP; they are also known as Complex V).

Here are some examples of rotors that you're probably familiar with. What makes these rotors spin? Wind turbines you pass on the highway are powered by the flow of air; an old watermill is powered by the flow of water; the same is true for modern hydropower plants; locomotives used to be powered by the flow of steam. ATP synthase is powered by the flow of positively charged particles called protons.

Recall that there are two separate areas in a mitochondrion: the intermembrane space and the matrix (see figure on page 18). The protons are positively charged, and they are attracted from the intermembrane space to the negatively charged mitochondrial matrix. The normal way they can go to the other side is through some specialized openings in ATP synthase. As the protons flow through these openings, they turn the rotors, which produce ATP.

The key point is that there is a difference in concentration of protons between the intermembrane space and the matrix. This difference is called the proton gradient, and the obvious question is: How is this proton gradient formed? There are special pumps that push these protons from the matrix to the intermembrane space. This process is called proton pumping.

PROTON PUMPING:
PRODUCTIVE BUT POTENTIALLY POLLUTING

Feel free to skip this box, but don't forget the tongue twister above: proton pumping is a productive process—it is what allows the rotor to spin—but it is also a potentially polluting process; this is how ROS are created. Proton pumping is almost as fascinating as the rotors spinning, and it occurs as a result of the electron transport chain. At one end of the chain, there are two molecules (NADH and $FADH_2$), each of which gives two electrons to the process, and at the other end of the chain there is oxygen, which accepts these electrons. Between the two ends of the process there are four complexes: Complex I, II, III, and IV that are embedded in the inner membrane. As electrons move through this chain, Complexes I, III, and IV spit protons (H^+) into the intermembrane space, creating the proton gradient. A by-product of an imperfect electron transport is pollution in the form of ROS.

How does all this relate to the food you eat and to the air you breathe? The sandwich you had for lunch and the oxygen you're inhaling right now enable the process of proton pumping. You may recall from chapter 1, two molecules that are the product of the Krebs cycle: NADH and $FADH_2$. These two molecules feed the proton pumping. The oxygen you breathe is needed at the end of this process.

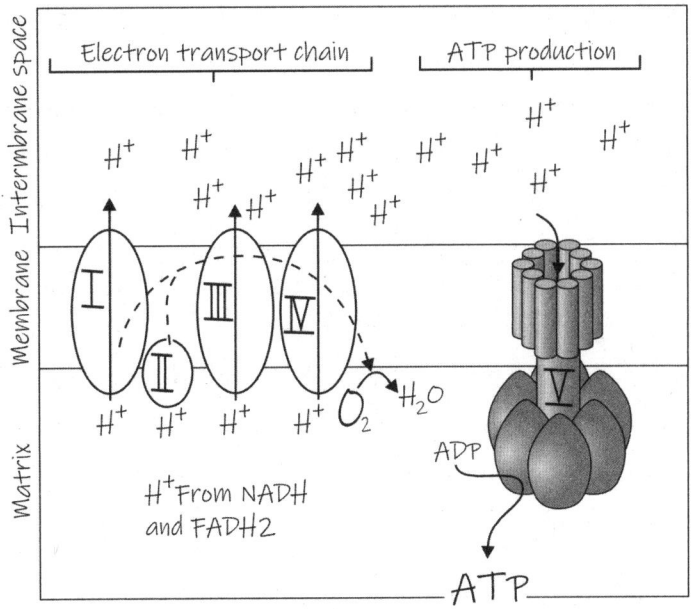

How the food you eat and the oxygen you breathe end up as ATP.
What you eat and breathe enables the process of proton pumping,
seen on the left side of the picture. Two molecules, NADH and $FADH_2$,
that have their origins in what you eat start the electron transport
chain (Complexes I to IV). The oxygen you breathe (O_2) accepts electrons at the end of the process. As electrons move through this chain
(represented by the dotted lines), positively charged particles called
protons (H^+) are pumped into the intermembrane space, creating the
proton gradient—a difference in concentration of protons between
the intermembrane space and the matrix. Because there is now a negative charge in the matrix, the protons are attracted to that area of the
mitochondrion, and the only way back is through the rotor, which can
be seen on the right side of this (simplified) diagram. This rotor (yes, it
literally spins around) is called ATP synthase, or Complex V, and it converts ADP into ATP, which energizes the cell activity.

Without oxygen or without a proton gradient (created through proton pumping), this process doesn't work. The rotor doesn't spin, and there is no ATP. No energy. No life. We wish we could shrink to the size of a mitochondrion to see those rotors with our own eyes. There is currently no equipment to see the spinning in action. This may be possible in the future, but until then, there are several online videos that show how this rotor works which are worth watching.[14] Three scientists were awarded the Nobel Prize for helping us understand all this. Some of this knowledge was refined through one of the most bitter scientific debates in twentieth-century biochemistry. It is known as the oxphos wars because the process we described is called oxidative phosphorylation.

HISTORICAL PERSPECTIVE: THE OXPHOS WARS

At the height of this debate in the early 1960s, one camp consisted of prominent scientists from all around the world, and the other camp consisted of two people: Peter Mitchell and Jennifer Moyle. Mitchell was one of the most colorful characters who ever operated in the field. Picture a graduate student at Cambridge who looked like Beethoven driving his Rolls-Royce around town.[15] He was a man of big ideas, and luckily for him, he had Jennifer Moyle as a partner. Franklin Harold, a scientist who collaborated with the two, believes that without her, Mitchell's imaginative theory would probably have ended in oblivion.[16] Being a true iconoclast, Mitchell didn't find his place in academia, so he and Moyle conducted their research from a private research institute (financed by Mitchell and his family) in a remote mansion in South West England.

The debate between the two camps was about how ATP is generated. It was already known that most ATP is produced inside the mitochondria past the Krebs cycle, which we described in chapter 1. The big question was: How? The scientists who belonged to the majority camp (Bill Slater from the Netherlands, Britton Chance from the United States, and others) were on a hunt for a molecule that transfers the energy derived from respiration into ATP. (This followed discoveries by Fritz Lipmann, who had shared the Nobel Prize with Hans Krebs.) Peter Mitchell was the first to suspect that there was no such molecule, and he came up with the radical hypothesis of a proton gradient. It was already known that a mitochondrion has two membranes—an inner membrane and an outer membrane, and Mitchell believed that what drives ATP synthesis is the charge difference across the inner membrane.

When presenting his ideas at scientific conferences in the early years, Mitchell faced a lot of blank stares and rolled eyes. And not surprisingly, the fiercest opposition came from the scientists who were searching for the elusive molecule that would explain everything. Scientific debates can get personal and even nasty, and the debate with Mitchell had its heated moments. At a 1966 conference in Warsaw, Mitchell was seen in an argument with one of his main opponents, Bill Slater, who was so furious that he began hopping on one foot. (The British biochemist Brian Chappell who recalled this commented that he finally understood the meaning of the expression "hopping mad").[17] But over time, Mitchell, Moyle, and others presented evidence that, indeed, ATP production is powered by a proton gradient.[18] In 1978, Peter Mitchell was awarded the Nobel Prize in Chemistry, and it is likely that if it happened today, he would have shared it with Jennifer Moyle. For their work on Complex V (ATP synthase, that amazing molecular rotor), Paul D. Boyer and John Walker won the Nobel Prize in 1997.

How does all this relate to aging? During proton pumping, this sophisticated machine creates pollution in the form of ROS. When electrons move between complexes, some of them "leak" and form reactive oxygen species

(ROS), which are a class of free radicals—compounds that have free electrons looking to pair with another atom and can wreak havoc in the process. (Earlier, we referred to them as pollution, but you can think of free radicals as troublemakers that bring chaos to whatever social group they join.) Most free radicals in our body are created inside our mitochondria as part of the energy transfer process. The good news is that mitochondria have a feature that parents of most teenagers would love for their kids to adopt: they clean up after themselves. Mitochondria don't only generate ROS but also manufacture antioxidants that neutralize those ROS. But when there's too much ROS accumulation, the antioxidants can't neutralize the reactive oxygen species fast enough, a phenomenon known as oxidative stress, which can be defined as an imbalance between oxidants and antioxidants in favor of the oxidants.[19] *Oxidative stress* is an important term that will appear throughout the book and not only in the context of aging, but it is common as we age because the cleanup process becomes less effective over the years. Like any other machine, the life machines create pollution, in this case, in the form of ROS, which can damage the cell in three key ways.

First, they can break other components of the cell, which would then need to be replaced, a task requiring a lot of ATP that would now be diverted from other tasks. Under this condition of oxidative stress, mitochondria are not unlike a nuclear power plant that used to produce clean energy until it broke down, and now emits pollutants that destroy everything in sight. Further, when free radicals interact with the building blocks of the cell, they create aldehydes, a class of molecules that adhere to other molecules permanently, and prevent their flexibility, mobility, and normal functions. You can think of aldehydes as rust, and over the years, we accumulate more and more of this "rust" in our mitochondria. Again, there's good news: There are several enzymes in the mitochondria in charge of disposing of those aldehydes, but the "rust" accumulates if those enzymes can't do it fast enough. And this "rust" can spread to other parts of the cell and to other tissues and create further damage. An example of the damage created by aldehydes is found in some neurodegenerative diseases, which are characterized by accumulation of protein aggregates in the brain. Typically, these are formed after chemical binding of aldehydes, especially aldehydes derived from interaction of ROS with mitochondrial fatty acids. (Smoking, vaping, and drinking alcohol are among the things that generate even more aldehydes. More on this in chapter 11.)

Now let's move on to the second way in which ROS can injure the cell. Many researchers believe that this is how mtDNA becomes damaged over time. Nuclear DNA is protected by proteins in the same way you protect a fragile item with bubble wrap. Unlike nuclear DNA, mtDNA is quite "naked," making it more vulnerable to mutations relative to nuclear DNA. Also, unlike nuclear DNA, the machinery inside the mitochondria to correct mtDNA mutations is much more limited. After years of operation, those circular mtDNA have accumulated mistakes during repetitive replication and have also been hit by ROS, partially as collateral damage of regular activity, and partially from exposure to pollutants. We described in the previous chapter the awesome pruning process of mitophagy by which damaged mitochondrial parts, including damaged mtDNA, are destined to removal. However, over time, as more and more damaged mitochondria accumulate, the mitophagy machinery becomes exhausted, too. In addition, mitophagy doesn't come cheap; it requires ATP as well, which is harder to come by as we grow older, since the life machines (mitochondria) are faltering.

Third, although nuclear DNA is better protected than mtDNA, it, too, can be damaged by ROS and aldehydes. Your genes are made from DNA, and they serve as the instructions for building your body and maintaining it. So if these reactive oxidative species and aldehydes hit nuclear DNA, they can damage it, which is like tearing a page out of the instruction manual for all the cell's functions.

The link between ROS damage and aging has been known as "the free radical theory of aging" and it has been around for several decades. In recent years, the theory has been questioned by several researchers, for example, by those who argue that the damage or mutations in the mtDNA are due to replication errors, or those who highlight other aspects of mitochondrial dysfunctions that are associated with aging. It is beyond the scope of our book to describe the different nuances of this debate; however, the most important takeaway is that mitochondria deteriorate over time, and this deterioration is associated with aging.

"I Intend to Live Forever. So Far So Good."

In the 1990s, a researcher named Nir Barzilai wanted to understand why certain people live to be one hundred, and he recruited several hundred Ashkenazi Jews around the age of one hundred who lived independently. Barzilai, who is the director of the Institute for Aging Research at Albert Einstein College of Medicine in New York, thought that maybe these folks live longer because they have healthier habits than the general population, but what he found surprised everyone: Almost half of them were overweight or obese, nearly 50 percent were smokers, and fewer than 50 percent did even moderate exercise. One of his study participants was Helen Reichert, who smoked for more than ninety years. When Barzilai asked Helen, "Didn't any of your doctors tell you to stop smoking?" she said, "Sure, but all four of those doctors died."[20] His centenarians study grabbed the media's attention, but when we interviewed him, Barzilai pointed out that many in the media missed the point. "The media made it sound like we're saying that exercise and diet don't matter," he said, when in fact the main takeaway was that these centenarians were somehow being protected by genetic differences that the general population doesn't have. He emphasized that even for people who may be protected by longevity genes, he recommends a healthy diet and regular physical activity, and he certainly doesn't recommend smoking. Another lesson from Barzilai's research relates to humanin, the mitochondrial-derived peptide that we mentioned in chapter 2. Together with other scientists, including Hassy Cohen, Barzilai found that humanin (which has protective properties) is found at high levels in families of the centenarians in his study.[21]

Barzilai belongs to a growing group of scientists who believe that by treating primary processes that drive aging, they may be able to prevent (or at least delay) the onset of all age-related diseases. Scientists have identified twelve hallmarks of aging[22]:

1. Damage and mutations to DNA (genomic instability).
2. The protective ends of chromosomes get shorter (telomere attrition).
3. Changes in gene activity (epigenetic alterations).
4. Problems with folding proteins into the correct structure (reduced protein quality control).
5. Autophagy and mitophagy don't work as well (disabled macroautophagy).
6. Cells are less responsive to nutrients (deregulated nutrient-sensing).
7. Problems with mitochondria on multiple fronts (mitochondrial dysfunctions).
8. "Zombie cells" that stop growing or dividing but don't die (cellular senescence).
9. Less ability to repair tissues and replace the cells that died (stem cell exhaustion).
10. Reduced communication between cells (altered intercellular communication).
11. Low-grade inflammation without overt infection (chronic inflammation).
12. Imbalance in the microbiome (dysbiosis).

Without going too deep into these, we want to make three points. First, you'll notice that mitochondrial dysfunction is one of the hallmarks of aging. Second, there is plenty of research regarding the links between the different hallmarks. For example, we already pointed out that attrition of telomeres leads to mitochondrial dysfunctions, and vice versa. The third point has to do with health span, and here is how Barzilai puts it: "The important thing is that you don't have to fix all of the hallmarks of aging to have an increase in health span," he told us. "You can target the mitochondria itself, and you'll have less inflammation, better metabolic regulation, better stem cell function, and so on. Improving one of these processes will frequently benefit the others."

Trying to stretch health span is of course different from aiming for immortality. (Health span refers to the number of years one lives free of disease, as opposed to lifespan, which refers to how many years one lives, and immortality, which is living forever.) In recent years, there's been serious discussion in scientific circles about extending human life well beyond one hundred years. Some even believe that we can achieve immortality (which is why we chose comedian Steven Wright's one-liner as the title of this section). While there is certainly some hype in the claims about longevity, the hope to extend health span is not unfounded since it is becoming clear that dysfunctional mitochondria play a role in many diseases. If we enhance mitochondrial functions, we'll give the cell the energy to fix the problem, and it is quite possible that all diseases—chronic and acute—can benefit from healthier, better-functioning mitochondria. So although we certainly don't expect to live forever, the leaps in mitochondria research increase the likelihood of extended lifespan and healthy aging as treatments for more diseases will be found.

•   •   •

Back to our hike in Kumano Kodo. Legend has it that six hundred years ago a Japanese princess named Terute carried the sick samurai she was madly in love with all the way to Yunomine Onsen, the remote hot spring village where we stayed. The man, Oguri Hangan, was blind, deaf, and couldn't talk or walk, so she carried him for three days and three nights in a cart. She had him soaked in the sacred water of the bath every day. After seven days he opened his eyes, after two weeks he could hear again, after three weeks he was able to talk, and after forty-nine days he was cured of all his ailments and was a young man again.

The day after our climb in the dark, we soaked in this same ancient hot bath. This was before we thought of writing this book, so we weren't thinking about our mitochondria. We knew, of course, about the therapeutic effects of hydrogen sulfide from hot springs that had been recognized since ancient times by the Romans and others. While researching this book, we learned that at least part of these therapeutic effects can be attributed to mitochondria. Hydrogen sulfide that is given in small amounts improves mitochondrial functions, regulates components of the Krebs cycle, and protects mtDNA from damage. Researchers from the University of Exeter in the U.K. increased hydrogen sulfide in the mitochondria of small worms and learned that it extended their health span, but not their lifespan. As with anything else, excess can be toxic, and inhaling too much hydrogen sulfide can be fatal. From our nonscientific experiment with this ancient bath, we can say that it was lots of fun, we think we noticed that our skin was glowing, and that we slept very well that night.

The advice we mentioned in the previous two chapters is relevant to aging, too. Sleep requirements may decrease a little over the years, but getting a good night's sleep is essential for mitochondrial functions in old age, too. And as we'll explain in chapter 6, exercise is particularly important in our later years because muscle loss (sarcopenia) in the elderly is a serious problem that can often lead to disability. And remember, aging does not begin when you retire, and while it's never too late to start, research indicates that those who start exercising early have an advantage.

After two blissful nights in Yunomine Onsen, we were sad to leave, but in the morning, we packed and moved on. We loved Kumano Kodo so much that in 2019 we visited the area again, including that problematic trail (in full daylight this time). We couldn't believe that we walked it in the dark without breaking a single bone. We try to be better now about planning our hikes well in advance. And as for the last stretch of our lives—the years we have left on this planet—all we can do is try to do the right things: exercise, walk, eat healthily, sleep well, not stress too much. In short, we try to be nice to our mitochondria and hope that they'll be nice to us, too.

CHAPTER 4

Engine Malfunctions

On July 16, 1982, Dr. William Langston got an urgent call from a young physician at his hospital. As the director of neurology at the Santa Clara Valley Medical Center in San Jose, California, he was routinely called to help with diagnosis, but what he saw when he arrived at the patient's room was baffling. A man was propped up in the hospital bed staring straight ahead and drooling. The man was clearly awake, but he was eerily motionless, as if frozen.[1] To Langston, it looked like a classic case of advanced Parkinson's disease, which causes loss of motor control, except for two puzzling observations: the man was forty-two years old, which is typically too young for Parkinson's, and more importantly, his symptoms appeared in a matter of days, which doesn't happen with Parkinson's, where such symptoms develop gradually over years.

As we'll see in a moment, this puzzling case led to a breakthrough in establishing the link between mitochondria and Parkinson's disease, and in this chapter, we'll discuss a range of diseases involving impaired mitochondria: from neurodegenerative diseases, to type 2 diabetes, cancer, and more. In some of these cases, scientists believe that mitochondrial dysfunctions cause the pathology. In other cases, they suspect that the ailment causes the mitochondrial abnormality, and in many cases, they simply don't know which comes first. This is important: We don't claim that malfunctioning mitochondria are the cause of all diseases, but that many are accompanied by mitochondrial dysfunctions that make things worse.

As we saw in chapter 1, mitochondria are critically important for many processes beyond energy transfer: providing building blocks, fighting viruses, getting rid of cells that should die, signaling . . . all these and other functions can be impaired. And dysfunctional mitochondria further add directly to cell injury by increasing oxidative stress—the imbalance of free radicals and antioxidants, which we described in chapter 3. Understanding how mitochondrial dysfunctions contribute to different diseases may lead to new interventions and treatments. By improving mitochondrial functions, we may reduce the likelihood of certain diseases, delay their onset, or decrease their severity. In the next chapter we'll explain how this can be achieved through medical interventions, but first, let's dive deeper into the links between mitochondria and some common and less common diseases.

Neurodegenerative Diseases

Today, there is overwhelming evidence linking Alzheimer's, Parkinson's, and several other neurodegenerative diseases with mitochondrial dysfunctions.[2] This wasn't always the case, and the story of the frozen patient was a breakthrough in this space. When William Langston met his patient, the man couldn't move or talk. Then, through painfully slow communication, the doctors figured out that trapped inside this frozen body was a person who could hear and feel everything. His symptoms appeared two days after he and his girlfriend tried a new synthetic heroin, and before long, the girlfriend and five additional cases were located. They were all in Northern California, all young and frozen, and all used the same synthetic heroin. Local and national authorities got involved in trying to solve the puzzle, and they made a significant leap when someone remembered a case of a patient who exhibited similar symptoms after he took a drug that contained a compound called MPTP.[3] Lab results of the heroin taken by Langston's patients indicated that it also contained MPTP, and that led to an explosion of research clarifying the link between Parkinson's disease and mitochondrial dysfunctions.

Is Parkinson's caused by MPTP? No. But the fact that MPTP causes Parkinson-like symptoms allowed scientists to create an animal model of the disease. (Animal models mimic the symptoms of the disease in question.) After Langston reported the case of the frozen patients in the journal *Science*,[4] researchers from all over the world immediately realized the opportunity to test different hypotheses and interventions; by giving MPTP to mice, they could create Parkinson's symptoms and test their ideas. The big insights from the studies that followed were, first, that MPTP starts a cascade of events that is toxic to Complex I in the mitochondria. Second, that some common pesticides may cause Parkinson's disease by the same mechanism (more on this in chapter 11). Third, and most important, that a mitochondrial defect is sufficient to cause Parkinson's disease.

We now know that mitochondrial dysfunctions are found in other common neurodegenerative diseases, including Alzheimer's. A partial explanation to the close link of neurodegenerative diseases and mitochondria is the high energy demand of the brain. Although your brain represents around 2 percent of your body weight, it accounts for about 20 percent of your energy use, so when your brain doesn't get enough ATP, this means trouble. But since mitochondria do much more than energy transfer, it isn't only about reduced ATP production.

Mitochondrial dysfunctions trigger neuroinflammation (inflammation in the brain) typical of all neurodegenerative diseases. When the mitochondria in a neuron are damaged, they release their components into the fluid part of the neuron (the cytosol), including mtDNA. These mitochondrial components are detected by the same system that detects viral and bacterial invasions, to express cytokines, the signaling molecules of inflammation, leading to innate immune response. This response can become chronic by activating other brain cells called glia. Such chronic innate inflammation is damaging to the neurons; it can also disrupt the blood–brain barrier (which protects the brain from toxic compounds as well as bacteria and viruses). The sustained inflammation and the breakdown of the blood–brain barrier cause further pathology.

Neurodegenerative diseases have also been linked to problems with other functions of mitochondria: their role as metabolic hubs,[5] disrupted mitophagy,[6] excessive fission in that fission/fusion dance, and more. And dysfunctional mitochondria produce more free radicals than they can clean up, which causes oxidative stress. As you can see, there are many ways that link mitochondrial dysfunctions with neurodegenerative diseases, and these give scientists hope that treatments for those diseases may be found, in part, by protecting the mitochondria.

Type 2 Diabetes

Recall that one of the key functions of mitochondria is carbs and fat metabolism. Since type 2 diabetes is a classic metabolic disease, it's not surprising that it is so closely associated with mitochondrial dysfunctions, and to understand how, we need to introduce an important concept, "metabolic flexibility." As we mentioned, mitochondria use two main fuel sources that come from the food you eat: fatty acids (which come from breaking down fat) and glucose (which comes from breaking down carbohydrates). Mitochondria constantly switch between these two sources based on several factors. For example, the spike in blood sugar right after a meal triggers the mitochondria to use glucose as an energy source. Metabolic flexibility is the desired condition in which mitochondria can switch from glucose to fatty acid oxidation, when glucose levels are low. When mitochondria lose the ability to switch between the two sources, this is known as metabolic inflexibility, a condition that is found in people with obesity and type 2 diabetes.

Metabolic inflexibility is closely associated with insulin resistance, a state in which cells cannot efficiently pick up glucose from the blood; even when there is plenty of glucose in the blood, the machinery to "import" glucose into the cell is inefficient.[7] How does this manifest in people with diabetes? Chronic high levels of glucose in the blood can damage the eyes, kidneys, peripheral nerves, and more by reacting with and damaging proteins (a process called glycation). And when diabetics can't utilize fat as fuel, it accumulates and can lead to additional problems in their heart, liver, and other organs. Type 2 diabetics tend to have dysfunctional mitochondria,[8] and while there is still a debate on whether defects in mitochondria are the cause for or the result of type 2 diabetes, it's clear that improving mitochondrial functions can reduce the complications associated with the disease.

When we talked about signaling in chapter 1, we only gave examples of mitochondria sending signals, but mitochondria also *receive* them. For example, beta cells in the pancreas are known for producing insulin. A lesser known but critical part of their job is to sense the energetic status of the body and then tell the body what to do with its energy. Should it store it or use it? If so, where should the energy be used? In 2004, Orian Shirihai from UCLA showed that it is the mitochondria in those beta cells that do the work and explained that when something goes wrong with their sensing and signaling, the result is diabetes.[9]

Half of the adults in the United States are diabetic or prediabetic, and about a third are unaware of it.[10] While these conditions are common, we can fight them. Take Emanuel's experience as an example. In 2017, a routine blood test showed that the level of glucose in his blood was just above the normal level of up to 100 mg/dL. When Emanuel left our doctor's office after she delivered him the news, he felt defeated and disappointed with himself. He was exercising regularly; he watched what he was eating; he wasn't overweight (okay, maybe a little); and yet, Dr. F. had just given him the P word . . . prediabetes. Our family doctor for twenty years, Dr. F. was always cheerful, but she was serious, and she didn't smile even when Emanuel tried to lighten the situation by cracking a joke. He promised her that he would reverse those numbers, but he doubted it himself. The following month we went to Japan for that short sabbatical we mentioned earlier, and something interesting happened there.

When we arrived in Japan, the thought of driving on the left side of the road terrified us, so instead of renting a car, we ended up walking a lot. The house that we were staying in was on the top of a steep hill, so we got a lot of endurance exercise. Over the three months in Japan, we cut down on calories and ate mainly fish. By the end of our stay, we discovered that Emanuel had lost six pounds, and he had lowered his glucose level to the normal range. That appointment with Dr. F. was a real wake-up call that taught us two lessons: First, you don't need to be overweight to get diabetes. Second, at its early stages, diabetes is completely reversible. In fact, studies show that exercise can significantly improve the situation and is an effective treatment for diabetes in people who had been sedentary before.[11] An important development in this area is the availability of continuous glucose monitors, which can teach you about how your body reacts to different types of food or exercise. While still relatively expensive, these monitors are gaining popularity not only among diabetics.

Heart Diseases

Imagine that you had to pump two thousand gallons of water using a hand pump every day (just the thought makes our arms sore). Your heart knows what that feels like—it pumps that amount of blood 24/7 for 365 days each year. It's no wonder that, as mentioned, 40 percent of each heart muscle cell is made up of mitochondria,[12] and when those mitochondria are not in top shape, the heart's performance suffers.[13] Here's a look at what happens in heart cells of a patient (let's call her Sylvie) during and after a heart attack.

One day, Sylvie collapsed on the sidewalk holding her chest. She was sweating and breathed heavily; she was frightened. As she waited there for the ambulance to arrive, let's zoom into Sylvie's heart. During myocardial infarction, which is the medical term for a heart attack, part of the heart muscle (myocardium) experiences insufficient blood supply (infarction), which means that oxygen doesn't reach some cells.

WHAT HAPPENS IN ONE MITOCHONDRION DURING A HEART ATTACK?

Looking at a single mitochondrion in one of those cells, we would see how this machine comes to a halt. At first, electrons keep going through the electron transport chain, but when they reach the final point, they need to bind with oxygen, which now is impossible to find. With no oxygen, the electrons stop moving, which means that no protons move to the intermembrane space to create the gradient that powers the rotors. When the rotors stop spinning, they no longer produce ATP. This situation can cause the cell to simply stop in its track and cease to contribute to the cardiac pumping activity. If prolonged, such oxygen-starved mitochondria signal to the cell that it is time to die. This activates programmed cell death, and that one muscle cell in Sylvie's heart is gone forever. Other cells in the oxygen-starving area die the same way, or in a less elegant manner (through a process called necrosis), and their content spills and damages neighboring cells. It's a disaster zone with multiple dead muscle cells.

The ambulance arrived. Forty minutes later, at the emergency room, a balloon catheter was inserted through Sylvie's groin to the blocked blood vessel in her heart. The balloon was inflated for a few seconds to push the blockage against the artery walls and then deflated. Blood was flowing freely again, which is good news, but too much of a good thing can mean trouble sometimes. In the same way that a big meal can be fatal for a starved person, a sudden flow of oxygen into starved cells can mean trouble because a lot of reactive oxygen species (ROS) are created in the process and can damage or kill the cell. This phenomenon is known as reperfusion injury, and looking at these cells closely, you would see lots of damaged mitochondria that are not as neatly organized as before the heart attack.[14]

Sylvie's life was saved, but at that moment, she also may have joined a club that no one wants to belong to. Heart failure is a condition in which the damaged heart muscle doesn't work well, and as new interventions and medications improve the survival rates following a prolonged heart attack, some people like Sylvie find themselves in a growing group of patients who suffer from this condition. Two years after her heart attack, we may find Sylvie in an armchair in her living room, her legs swollen, huffing and puffing when she gets up from her chair. Her heart muscle may have healthy cells that operate as usual, but there are fewer of them, and they have to work extra hard, and in the process, they exhaust and damage their own mitochondria.

There are other ways in which dysfunctional mitochondria in the heart are part of the vicious cycle that contributes to heart failure in the aftermath of a heart attack. One of the problems involves calcium. Everybody knows about the importance of calcium for the strength of our bones, but it is also present in all tissues, and it plays a key role in signaling within the cell. One of the many tasks that mitochondria carry on their tiny shoulders is to keep calcium away from certain enzymes in the cytoplasm, because it can damage or even kill the cell. Known as calcium sequestration, you can think of it as crowd management in a sports arena hosting a game between two notoriously rival teams. Fans of the two teams are assigned to separate sections; otherwise, there will be blood. However, there are times (e.g., during halftime or the end of the game) when fans from both teams will mix up, and you want to make sure these are managed well.

Back to what happens in our heart: Every contraction is accompanied with a rise and a quick decline of calcium in the cytoplasm of the muscle cells because calcium is the way the muscle gets the message that it should contract. When contraction is less regulated and calcium levels stay high, the mitochondria try to help by taking it up. But calcium overload in the mitochondria is damaging to the mitochondria themselves, causing ROS production and further cardiac damage. You can think of it as the arena staff pushing too many fans of one team into a small section, which often doesn't end well.

The bottom line for Sylvie is that her damaged heart muscle continues to deteriorate. Eventually, the muscle may give up, and as a woman, Sylvie is at a disadvantage, according to a 2020 study from the University of Alberta in Canada. The study indicates that women face a 20 percent increased risk of developing heart failure or dying within five years after their first severe heart attack compared with men.[15] Cardiovascular diseases are a leading cause of death around the world. Many of the lifestyle practices that have been proven to be good for your mitochondria are also good for your heart and especially after a heart attack: eating a healthy diet, maintaining a healthy weight, giving up smoking, reducing alcohol consumption, increasing physical activity (per advice of your physician), and more. Keeping your blood pressure under control is another important one because high blood pressure can damage the elasticity of your arteries, which might decrease blood flow to your heart. At the cellular level, research shows that high blood pressure can degrade mitochondrial function.[16]

Psychiatric Disorders

There is mounting evidence that associates psychiatric disorders with mitochondrial dysfunctions, but the idea still faces some resistance. Carmen Sandi, a professor at the Swiss Federal Institute of Technology in Lausanne, Switzerland, remembers the first time she presented her group's research on mitochondria. Some prominent neuroscientists who sat in the front row were shaking their heads as she presented, and after her talk they pulled her aside and told her that she was wrong. The experience repeated itself when she presented her findings at the 2015 British Neuroscience Association meeting in Edinburgh. "After that I was very stressed every time I had to present, I developed a kind of PTSD," she says half jokingly. As someone who has been studying stress and anxiety, she knows a thing or two about the subject. What was so controversial in her talk wasn't the idea that mitochondria change in the brain under certain conditions, but that they are the primary drivers of the change. In other words, she argued that depression and anxiety are not necessarily driven by changes in neuron-to-neuron communications that affect the mitochondria but the other way around: The changes in the mitochondria impair communications between neurons. This may mean that by treating the mitochondria we can decrease the burden of psychiatric disorders, an idea that still goes against the dogma in the field.

Proving causality, not just correlation, is one of the toughest challenges in science, and Sandi and her team were able to demonstrate (in rats so far) that correction of mitochondrial dysfunction is sufficient to decrease depression and anxiety.

They found that highly anxious animals showed increased depression-like behavior, and that their mitochondria were different, too. In the more anxious rats, the mitochondria in a certain area of the brain regulating reward and emotionality were different from normal ones in several ways. For example, the team found that in that subset of neurons the mitochondria had a lower level of a protein called mitofusin-2, which is important for fusion. And now to the kicker: To begin establishing causation, when Sandi's team raised the level of mitofusin-2 in these neurons, the anxiety-like and depression-like behaviors almost disappeared. Those rats that previously exhibited anxious behavior (for example, by avoiding open spaces) behaved more like non-anxious rats. Conversely, reducing mitofusin-2 levels in the same neurons of normally non-anxious animals (in this case they used mice) resulted in increased anxiety and depression-like behaviors.[17]

There are many clinical studies that demonstrate a correlation between anxiety or depression and mitochondrial dysfunctions in humans. For example, an international group of researchers led by Jonathan Flint, a behavior geneticist from Oxford University, looked for evidence of mitochondrial abnormalities in 11,670 Chinese women with and without recurrent major depressive disorder and found an abnormally higher count of mtDNA in the saliva of the women with major depressive disorder.[18] Considering what we discussed in the last chapter about the link between mitochondria and telomere length, the following may not come as a surprise: The Chinese women who suffered from major depressive disorder also had shorter telomeres. These researchers also found an abnormally higher count of mtDNA and shorter telomeres in mice exposed to chronic stress. Importantly, efficiency of ATP production in the livers of these mice was reduced in response to stress. Together, these studies in human and mice indicate a strong correlation between mitochondrial dysfunctions and depression. Correlation does not prove that mitochondrial dysfunctions cause depression or vice versa, but as we just saw in Carmen Sandi's studies, there are initial indications that mitochondria might be the drivers.[19, 20]

While the key role of mitochondria in psychiatric disorders is not yet widely accepted, it is gaining ground. More researchers recognize that a lot of what's going on in our brain has to do with metabolism, and since mitochondria are metabolic hubs, their central role is clear. Although the jury is still out about this, in his book *Brain Energy*, Harvard psychiatrist and researcher Christopher Palmer put it this way: "Mental disorders—all of them—are metabolic disorders of the brain," he wrote. And he didn't stop there: "All the symptoms of mental disorders can be tied directly to metabolism, or more specifically, mitochondria, which are the master regulators of metabolism."[21]

Cancers

The difference between a regular cell and a cancer cell is not unlike the difference between a mensch and an egocentric jerk. While a cell in normal tissue is constantly working the balance between its own needs and the needs of the community of cells it belongs to, cancer cells couldn't care less, and they employ an arsenal of strategies to grow with total disregard for the signals from the tissue around them. Like members of a crime organization, they're obsessed with expanding their territory—they steal, they push, and they outsmart other cells, which makes them a formidable enemy. Some of their sneaky asocial behaviors become possible by suppressing or bypassing certain mitochondrial functions. One such function is programmed cell death. A regular cell cares so much about its surrounding cells, that under certain circumstances it would kill itself to prevent damage to the tissue it is part of. As we discussed earlier, in cancer cells, this mechanism is broken, so cancerous cells that should die continue to proliferate.

Cancer cells are sneaky about ATP production, too. As early as 1924, a German scientist named Otto Warburg observed that instead of relying on their mitochondria to make ATP, cancer cells use a process called glycolysis, which doesn't require oxygen.[22] This finding was puzzling because glycolysis is a very inefficient way to produce ATP. While glycolysis produces only four molecules of ATP from each molecule of glucose, mitochondria produce thirty

to thirty-two molecules of ATP. Why do tumor cells use less efficient ATP production? Warburg believed that cancer is caused by mitochondrial dysfunction
and that this is the reason cancer cells switch to the less efficient glycolysis.
Today, most scientists believe that he was wrong about this point. The answer
seems to be rooted in the speed of ATP generation in glycolysis, which better
matches the energy demands of the rapidly proliferating cancer cells; glycolysis
is ten to one hundred times faster than mitochondrial respiration in producing
ATP.[23] Essentially, cancer cells are wasteful but fast; they extract less energy
from each molecule of sugar, but they do it at a speed that enables their rapid
growth. This strategy works for cancer cells for another reason: Blood vessels
don't develop fast enough to feed the tumor, which means that there is less oxygen, but because glycolysis doesn't require oxygen, tumor cells can still produce
enough ATP to continue to grow. Cancer cells also have lower ROS production
and thus less ROS-induced damage because the mitochondria are much less
active—another benefit of switching to glycolysis. While Warburg was wrong
about seeing cancer as driven by mitochondrial dysfunction (a reminder that
correlation doesn't mean causation), his observation that cancer cells use glycolysis to generate ATP has been confirmed and is called the Warburg effect.

Cancer cells are inconsiderate of other cells also when it comes to the use
of nutrients. They steal nutrients and other molecules from other cells, even
though they still use the Krebs cycle in their own mitochondria to generate
building blocks for themselves. If this is not outrageous enough, they even steal
intact mitochondria from other cells to support their survival. Shiladitya Sengupta, a researcher from Harvard Medical School, was working on cancer, and
he kept seeing cancer cells that were not affected by the immune cells (called T
cells) that come to fight them. In 2022, using sophisticated microscopy tools,
Sengupta and his team were able to observe cancer cells that send out very fine
nanotubes to connect with immune cells. In further experiments, they were
able to demonstrate what is transported in those nanotubes—mitochondria.
Those sneaky cancer cells literally suck the life out of the cells that come to
fight them![24] Like an army that captured the enemy's supply, this siphoning of
mitochondria by the cancer cells weakens the immune attack of the T cells and

gives the cancer cells even more building blocks and energy. There's more: In 2025, researchers from Japan found that cancer cells have two more nasty tricks in their bag. First, they "poison" the immune cells by transferring their own damaged mitochondria into those immune cells. Second (and even nastier), the cancer cells provide these damaged mitochondria with a "cloak of invisibility" hiding the "poison" from removal by mitophagy in the immune cells.[25]

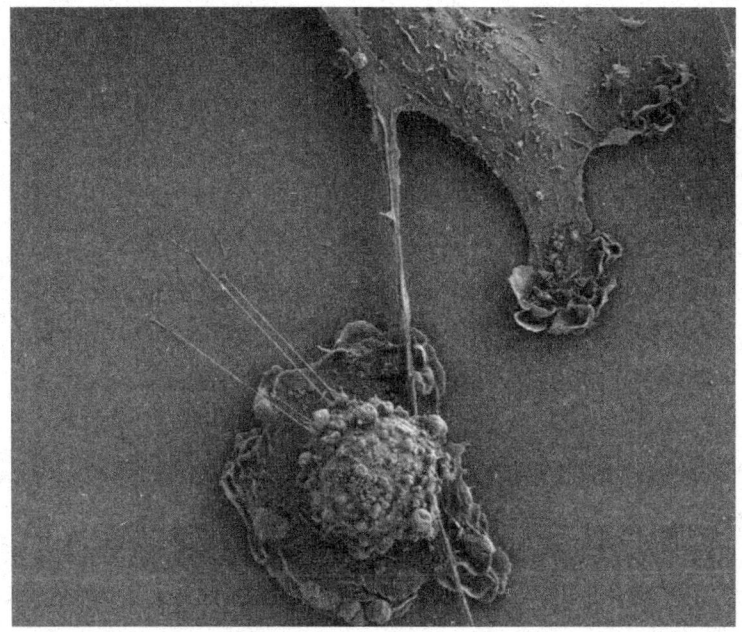

Caught red-handed. Cancer cells steal mitochondria from immune cells (T cells) that come to attack them. They do it through tiny "straws" called nanotubes that extend toward the T cells. In this electron microscopy image, the cancer cell is seen in the upper part and the T cell in the lower part. (Courtesy, Shiladitya Sengupta, Harvard Medical School)

Yet in the end, sociopaths can't hide because they identify themselves through their behavior. Similarly, the unique asocial behaviors of cancer cells provide ways to identify them. For example, Otto Warburg observed that cancer cells consume enormous amounts of glucose, and this feature is being used to this day in scanning for cancers; to identify where cancer cells hide in the body, PET scan machines are looking for locations with high glucose uptake. Here's another example of unusual behavior that draws attention: After a bank robbery, a vehicle zigzagging through traffic may help police identify it as the getaway car, and researchers at Harvard Medical School suspect that similar zigzagging may be going on when cancer starts to metastasize from one organ to another. As you recall, mitochondria use mainly fatty acids or glucose to feed the Krebs cycle, but research led by Marcia Haigis, a professor at Harvard, demonstrated that in the same person, the metabolites of a tumor change when the cancer moves to a different organ. For example, they showed that to sustain their Krebs cycle, breast primary tumors use glucose and glutamine, but that lung metastatic cells from these tumors use pyruvate, and brain metastatic cells use yet other substrates. Haigis and others believe that this behavior may help develop new strategies to identify and fight metastatic cancers.[26]

Cancers are complex diseases with a variety of tricks to fight anticancer cell defense mechanisms and our immune response. But again, all these tricks that they use to enable their fast growth and unique metabolism are also their Achilles' heel, which gives hope for new treatments.[27]

Autism Spectrum Disorder

It's important to note that not everyone sees autism as a disorder; some people on the autism spectrum and their families do not wish to change it but rather to build awareness and acceptance of neurodiversity. Yet many still hope that the medical community will develop treatments for this condition, especially in severe cases of self-harming behavior and delayed or absence of language skills. Some hope stems from the fact that mitochondrial dysfunctions are found in much higher percentages in children with autism spectrum disorder than in kids who don't have this condition.[28] Finding the cause of a disorder holds great potential for therapy, yet the intersection of three highly complex topics—

autism, mitochondria, and brain development—is not likely to result with one simple answer that will explain the many phenomena on the spectrum.[29]

Doug Wallace, whom we met in the previous chapter, is now the director of the Center for Mitochondrial and Epigenomic Medicine at Children's Hospital of Philadelphia. He sees a causal relationship between mitochondrial dysfunctions and autism. Wallace and his team developed mice with a mutation in their mtDNA and demonstrated that this mutation led to impaired social interaction, compulsive behavior, and increased anxiety in the mice.[30] As we discussed earlier, it isn't the first time in his long career that Wallace goes against the grain, and today several of his findings about mtDNA have been accepted. Time will tell if the same is true for his view on autism.

Robert Naviaux, a physician and a scientist from UC San Diego, believes that what occurs in people with autism can be explained through the theory of "cell danger response."[31] This response can be triggered by chemical, physical, infectious, or other biological threats to the cell. It is a normal mechanism that our body uses to initiate healing. Normally, in response to such threats, cells release ATP to their surroundings in proportion to the degree and duration of the threat, and this ATP builds up outside the cells and serves as an alarm system, triggering the necessary immune response and repair. After a few days, things usually get back to normal.

Naviaux believes that children at risk for autism are born with potential oversensitivity to changes in the cell's environment. When their brain is developing and if their neurons face persistent threats, these neurons continue to unnecessarily sound the alarm even when the danger has passed. In other words, their neurons overreact to even very small amounts of ATP outside them and keep behaving as if they are still in danger. The problem is that responding to damage uses a lot of energy, which requires the cell to devote precious ATP to this task rather than to other essential cellular tasks. Some examples of these important tasks include establishing proper nerve connections required for things like speaking and social behavior.

But what if we can stop the alarm system when it is no longer needed? Naviaux looked for a way to block this unnecessary signaling and chose a drug called suramin, which was developed in 1916 to treat African sleeping sickness. Following some studies in mice that supported his hypothesis, he then conducted a small clinical trial with ten boys, ages five to fourteen years, who had been diagnosed with autism spectrum disorder. Five of the boys got suramin, and five got the placebo, and real improvement was detected in the suramin group; kids who were hardly talking before were now saying complete sentences.[32] The benefit of suramin was tested independently in a clinical trial of fifty-two boys; and analysis of a subgroup of younger kids showed similar benefits,[33] supporting Naviaux's theory. These and other studies that implicate mitochondrial dysfunctions and altered metabolism in autism spectrum disorder led to additional mitochondrial-focused treatments, including the use of ketogenic diet and L-carnitine supplementation. Although the benefit of these interventions has not been proven, the mitochondrion is a novel treatment target that may improve the lives of children with autism spectrum disorder.[34] However, associations do not *prove* that mitochondria are the culprit in this syndrome, and more research is needed to support mitochondrial-focused therapy.

Chronic Fatigue Syndrome and Long Covid

Patients afflicted with these diseases often face skepticism about whether they are truly sick. If mitochondria could talk, they would tell a different story, and here is one case that illustrates it. When Paul Hwang and his colleagues at the National Institutes of Health zoomed in on Amanda Twinam's mitochondria, they saw a messy traffic jam. From tests they had conducted, they learned that her body was producing too much of a protein called WASF3, and now they witnessed how this protein jammed the ATP production machinery in her mitochondria. When Twinam heard the news from Hwang about the finding, she felt a sense of relief. For decades, she had been suffering from severe fatigue, and like many patients with chronic fatigue syndrome, she faced skepticism from the medical community. Chronic fatigue syndrome is an illness characterized by a wide range of symptoms, including severe fatigue, dizziness, pain, "brain fog," and exercise intolerance. According to the Center for Disease Control, at least 836,000 individuals in the United States (and possibly many more) suffer from this condition. Twinam, who also had cancer, compared society's attitude toward both illnesses. "Everybody believes you when you have cancer," she told the *Washington Post* in 2023, but this wasn't the case for chronic fatigue syndrome until she got the results from the NIH. "I can finally say, 'It's not psychological. I'm not a malingerer.' We now have a scientific explanation."[35] This is not necessarily a full or the only explanation for chronic fatigue syndrome, and it remains to be seen whether Paul Hwang's findings apply to a wide range of patients. When he and his colleagues checked muscle tissue of fourteen patients with chronic fatigue syndrome, they found an overabundance of the same WASF3 protein in nine of those patients.

The jury is still out on the question of which comes first, chronic fatigue syndrome or mitochondrial dysfunctions, and the same is true regarding the link with long Covid, which shares some similar symptoms.[36] A few months after the beginning of the Covid-19 pandemic, patients started to complain about lingering symptoms that were reminiscent of chronic fatigue syndrome, including extreme fatigue and foggy brain. Many of these patients faced initial skepticism, too, but after some time, studies validated their complaints, and some researchers suspect that mitochondria may be involved here, too. For example, when fifteen patients with long Covid were asked to participate in a study and perform low intensity exercise, they exhibited reduced mitochondrial functions relative to their performance prior to having Covid.[37, 38] One potential similarity with chronic fatigue syndrome is the link to a viral infection. Long Covid occurs after an infection by the virus SARS-CoV-2, and it is often reported that chronic fatigue syndrome also happens after a viral infection, which gives hope that finding treatment for one disease might help the other. Some scientists suspect that in both diseases, viruses hijack mitochondrial energy production and metabolism and inhibit other functions, including programmed cell death.[39] If true, this may explain the lingering symptoms. Recall from chapter 1 that part of the reason SARS-CoV-2 is so effective in spreading Covid is that it directly attacks the mitochondria-mediated immune response. So maybe the virus is even sneakier than had been thought before and uses mitochondria to its advantage in other ways. It's also possible that the similar symptoms of both diseases can be explained without mitochondrial involvement. Some researchers argue that these illnesses are due to organ damage caused by the viral infections.[40] Time will tell, and it's still a puzzle, but more and more clues point to a twelve-letter organelle that starts with an *M*.

Infertility

When Marisa Tomei stomped her high heels on the wooden porch and exclaimed "My biological clock is ticking" in the movie *My Cousin Vinny*, not much was known about the involvement of mitochondria in human reproduction. This was in 1992, but even today, mitochondria are missing from many discussions about fertility, yet there is growing evidence that they play an important role in the subject. For Vinny's biological clock, too! While aging men continue to produce sperm, the quality of their sperm declines, and it is possible that part of this decline derives from dysfunctional mitochondria that affect the speed and the direction of the sperm's movement. Like speedboats with faltering engines, the sperm advances toward the egg, but if they don't reach it fast enough, there will be no embryo. Why are older men more likely to experience infertility? Because over the years, as their primordial cells that make the sperm cells are exposed to radiation, nicotine, alcohol, and other harmful agents, these cells are at risk of getting mutations in their mitochondrial DNA and therefore give rise to defective sperm cells.

Infertility among women is widespread, too, and complications in getting pregnant and successfully completing the pregnancy are more common in older age. In an indirect way, one such complication led to a scientific breakthrough. In the late 1990s, a thirty-eight-year-old Turkish woman experienced complications that required frequent visits to the doctor. Her twelve-year-old daughter, Elvan, often came along to these visits, and while at the time Elvan was more interested in the Teenage Mutant Ninja Turtles than in medicine, those visits apparently made an impression. When she outgrew her childhood obsession with the green turtles, she became interested in some big open questions about human reproduction. In 2022, Elvan Böke, now the director of the Oocyte Biology Group at the Barcelona Center for Genomic Regulation, shed light on one of the biggest questions that puzzled scientists for decades.[41]

Every woman is born with the eggs (oocytes) that will eventually be used to create her children. Unlike sperm that are formed throughout the life of males, no new eggs are created during the woman's lifetime. Given what we learned in chapter 3 about ROS damage, how do these eggs maintain their pristine state? How can a cell that endured decades of ROS damage create new life? Elvan Böke and her colleagues found that in these eggs, the female body suppresses Complex I in the electron transport chain, which is part of the ATP production process in the mitochondria. Complex I is responsible for most of the ROS produced, and while suppressing it results in less efficient production of ATP, it also results in less ROS-induced damage. This is true in frogs and the same mechanism may be at work in humans. When and how Complex I becomes active again is not entirely clear, but it is possible that this finding may be the cause of some cases of unexplained infertility. Perhaps in some women's oocytes, Complex I starts forming earlier than is necessary, which means that it generates ROS (i.e., more damage) for a longer period, or Complex I is not activated in time, impeding successful fertilization. For females, there is some hope since studies show that these age-related problems could be treated, at least in part, by the administration of a supplement called CoQ10. More on that in chapter 7.

Primary Mitochondrial Diseases

At first sight, the lobby of the Sheraton Charlotte Hotel looked like any other lobby at a scientific conference: welcome signs, the reception table, name tags, the coffee station. But then, we started noticing them: a girl lying on a bench exhausted, her mother sitting by her side comforting her; a middle-aged man guiding his way with a white cane; a teenage boy in a wheelchair whose head and shoulders move uncontrollably left and right. The lobby was quiet, until we suddenly heard a roaring laughter from one of the corners. A group of teenagers was sitting around the table, some on chairs, some in wheelchairs. One thin boy was leaning on his walker. We had seen him earlier when he arrived by himself with the anxious expression of a transfer student on the first day of school. Now he had a wide smile on his face.

Unlike many other scientific meetings, the medical symposium of the United Mitochondrial Disease Foundation (UMDF) brings together scientists, physicians, and patients alike, so we were surrounded by the very people who would be the first to benefit from a breakthrough in mitochondria research. As we've shown throughout this chapter, many diseases and conditions are associated with mitochondrial dysfunctions, but what are primary mitochondrial diseases? The definitions vary, but they usually refer to a collection of rare, inherited conditions that impact the structure or functions of the mitochondria.[42]

Symptoms of primary mitochondrial diseases can show up in all stages of life, and the diseases can derive from mutations in mtDNA or in nuclear DNA. For example, in the case of Patricia and David Stallings, whom we discussed in the introduction, their boys Ryan and David Jr. (DJ) suffered from methylmalonic acidemia (MMA), which derives from a mutation in the nuclear DNA (only if both parents have this mutation, their child may suffer from MMA). In some other diseases, the mutation is in the mitochondrial DNA. For example, Kendall Conner, a seven-year-old girl, has such a mutation, which affects her central nervous system. Her mom, Taylor Conner, told Kendall's story from the podium at the opening session of the conference we attended in 2023. Nothing seemed to be abnormal with Kendall in the first two years of her life, and she met all the milestones. Then, a couple of months after she turned two, she started tripping and falling. Something seemed wrong, but her parents were reassured that things would resolve themselves. They didn't. After a second opinion, a third opinion, a neurologist appointment, an MRI, and genetic testing, Kendall's parents learned that she has a particular mutation in her mtDNA that leads to Leigh syndrome (which in some variations is caused by a mutation in the nuclear DNA). The syndrome is often caused by a deficiency in Complex I, which means that ATP production is seriously reduced. The impact on the brain (a big consumer of energy) explains the falls, the uncontrolled hand movements, and Kendall's speech difficulties. When her mom ended her presentation, Kendall approached the stage supported by her walker and her father, Rick. She walked up the steps by herself and stood next to her mother, proud and happy. While she has a hard time speaking, she's been an avid reader since she was four, and she wrote two children's books with her mom. The researchers who applauded Kendall and her parents don't necessarily study Leigh syndrome or other rare diseases, but many share the view that research of such rare diseases may lead to treatments for these patients as well as for patients with more common diseases. (For example, overcoming dysfunction of Complex I is also a possible treatment for Parkinson's disease.)

There is currently no cure for any mitochondrial disease. Richard Haas, a physician who's been treating such patients for decades, has seen some improvement in diagnosis. When he started working in the field in the late 1970s, there was no way to diagnose a disease like Leigh syndrome. Then, through the work on mtDNA by Doug Wallace and researchers like Anita Harding from the UK, who started to identify specific mtDNA mutations, diagnosis became somewhat easier. Yet the diagnosis is still pretty complex, and because fatigue is common and mitochondrial diseases present with symptoms in multiple systems, physicians like Haas are sometimes overwhelmed with patients who may suffer from other diseases. Unfortunately, when it comes to treatment, not much progress has been made yet. "From the therapeutic point of view, things haven't really changed very much, unfortunately, in almost fifty years," Haas told us. And yet, there was a sense of optimism at the UMDF meeting, a sense that something good may be just around the corner.

● ● ●

Another hint for the incredible importance of mitochondria in human health came in 2020 from an unexpected place: outer space. For astronauts, staying in space is hard on the body, and it seems to accelerate the aging of the heart, blood vessels, bones, and muscles.[43] An international team of researchers was looking for a possible underlying driver for these problems, and after analyzing data collected from fifty-nine astronauts, their results pointed in one direction: mitochondria. "We've found a universal mechanism that explains the kinds of changes we see to the body in space, and in a place we didn't expect," Afshin Beheshti, who led the project, said in a NASA publication. Referring to the effect of space travel on the human body, Beheshti added: "Everything gets thrown out of whack and it all starts with the mitochondria."[44] Many researchers agree that these issues would need to be addressed before humans could implement plans for space colonization or travel to Mars.

Yet as we have shown in this chapter, understanding mitochondrial dysfunction is highly relevant, even if you don't plan an intergalactic trip anytime soon. And the good news is that medical research related to mitochondria has come a long way; in fact, research in the field expanded exponentially to a point that the percentage of biomedical articles that mention mitochondria has surpassed all other organelles, including the nucleus.[45] We may be at the point when this theoretical knowledge is starting to be translated to actual solutions. There are currently dozens of clinical trials for mitochondrial dysfunctions and a sense of hope that is driven by years of discoveries in this field. We explore the emerging medical interventions for common and less common diseases in the next chapter.

CHAPTER 5

The Upside of Tending to Our Mitochondria

On the last day of February 2023, Ron Bartek was pacing back and forth in the National Institutes of Health hallways checking his phone for messages. A West Point graduate, Bartek had known some suspenseful moments in Vietnam and in his career at the CIA, but now he felt he was going to explode in anticipation of news from the Food and Drug Administration (FDA). Some twenty-five years earlier, Ron's stepson, Keith, was diagnosed with a rare neurodegenerative disease called Friedreich's ataxia (FA) that stems from a lack of frataxin, one of over a thousand different kinds of proteins that are found in the mitochondria. On this day, the FDA was supposed to announce its decision regarding a new drug called omaveloxolone to treat this disease. Bartek was at the NIH for Rare Disease Day, an event that the federal government holds once a year to raise awareness and discuss possible treatments for diseases like FA. He was hoping to hear the FDA's decision earlier in the day, but it was already after lunch and his phone was still silent.

Ron was introduced to FA when he and his wife, Raychel, noticed some lack of coordination, stumbling, and awkwardness in their young son, but they attributed those things to his fast growth. When Keith was eleven and things did not improve, Raychel took him to a neurologist, and after some initial tests, she called Ron from the clinic. She was in tears. She told him that the clinical examination led the neurologist to believe that Keith had an incurable disease called Friedreich's ataxia. When Raychel and Ron got home that evening, they started to search the web frantically, and the more information they found, the more they hoped that the neurologist was wrong. They learned that, if it were FA, over the years Keith would lose strength and coordination in his legs and arms; he would have an increased risk of diabetes; he'd most likely have severe scoliosis (sideways curvature of the spine); he'd lose much of his vision, hearing, and speech, and that he would most likely develop a very serious heart condition. When the results of Keith's blood test came in, Raychel and Ron's worst fear was confirmed. Keith had FA. In the months that followed, they found out that there was very little research going on about FA, and that there was no organization devoted entirely to this disease. A year later, they founded the Friedreich's Ataxia Research Alliance (FARA) with a bold goal: to cure Friedreich's ataxia. Over the decades that the organization has been in existence, it helped launch more than a dozen clinical trials, but no drug ever reached the finish line of FDA approval. While not a cure, omaveloxolone had been shown to slow the progression of the disease. But would the data provided to the FDA be sufficient to get approval? Would this be another disappointment, like the previous clinical trials, or would this be the time that they start to see an impact?

In the years following their son's diagnosis, Raychel and Ron saw Keith's condition deteriorate as his body weakened; holding on to someone's arm when he could no longer walk without assistance, then a wheelchair when he could no longer remain upright, a motorized wheelchair when his arms were unable to propel the manual chair. At one point, Keith got a tattoo across his chest: "How much time do we have?" He died at age twenty-four. Before Keith died, he thanked Ron and Raychel for doing all they could in hopes of developing treatments in time to help him. Knowing he had very little time left, he added: "That didn't quite work out but I know you'll keep going until you get it and you'll be in time for some of my buddies."

In this chapter, we'll discuss medical interventions aimed at helping the mitochondria in all of us, not only in patients with rare diseases. These cutting-edge developments will also help you appreciate why tending to mitochondria throughout your life is so important (the focus of part II of the book).

We can't change our genetic makeup, but by tending to our mitochondria, we do three things: First, we reduce the burden on injured cells, thus allowing them to heal. Second, we allow other cells to make up for the loss of damaged cells. And third, we can tip the balance in the cells in favor of good mitochondria. By doing this, we reduce chronic inflammation caused by the accumulation of broken mitochondria and cells in the tissue. Again, understanding how medical interventions work will deepen your understanding of the advantages you gain by making healthy lifestyle choices.

Reducing the Burden on the Cell, Especially as We Age

Friedreich's ataxia is a rare genetic disease, but the concept behind Omav (the nickname for omaveloxolone in the FA community) is an example that is relevant to the way common diseases like Alzheimer's and diabetes—and even aging—can be treated. By reducing the burden on the cellular repair machinery that addresses daily wear and tear, more disease-associated damages can be fixed. Mitochondrial dysfunction is not the cause of FA, and Omav doesn't cure FA. But Omav has been shown to reduce the burden on the cell by reducing oxidative stress, the condition where the antioxidants can't get rid of free radicals fast enough. By lowering oxidative stress, it reduces further damage that mitochondria bring about when they spin into a vicious cycle of creating more and more ROS.

While Ron was repeatedly checking his phone at the NIH, Ruth Acton, a mother of another FA patient, was at her home in Michigan, anxiously waiting for news about the same drug. With the drive of a desperate mother, Acton was the person who had sparked the process that led to the development of Omav. Her son, Jack, was diagnosed with FA in 2010, and she has been trying to learn about the disease ever since. After Jack would go to bed, she would spend many nights at her computer, trying to learn about possible treatments. While she's not a scientist, she had worked in finance at a pharmaceutical company, so she had a general idea on how drug development works, and she would often visit FARA's website, which listed past and present clinical trials and possible ways to treat the disease. One of these paths was reducing oxidative stress.

One day, Acton talked with another mother about a drug called idebenone, which was designed to reduce oxidative stress and was in a clinical trial for FA. This reminded her of a similar-sounding compound she had heard about from Jack's father, who is a chemist by training, with experience in the bio/pharma industry. That drug was under development for chronic kidney disease by a small Texas company, Reata Pharmaceuticals. Visiting Reata's website, Acton saw that they characterized their compound as an "antioxidant inflammation modulator," which made her think that maybe there was some possibility that their compound could belong to the antioxidant category that FARA and its partners were already evaluating. She also learned that Reata's pathway involved a protein called NRF2 that turns on protection from oxidative stress. Acton opened a new Google tab and searched NRF2 in Friedreich's ataxia and found a 2009 article by a French scientist named Pierre Rustin about the topic.[1] "Maybe this could lead to something," Acton thought. Is it possible that Reata could help FA patients? She checked with Jack's father, and he told her that it was worth pursuing. Acton didn't know anyone at FARA at the time, but the published research looked relevant, so she sent an email, and shortly after that got a call back from Raychel Bartek. This phone call triggered a series of events that eventually led to FARA and Reata joining forces and to the clinical trial for Omav.

The 2023 conference that Ron Bartek attended at the NIH was coming to an end. Joni Rutter, the director of the NIH center that hosted Rare Disease Day, stood at the podium making her final remarks. While she was talking, Ron got the call that he was waiting for. "I walked down the aisle," he told us, tears in his eyes, "looked up at the podium, and Joni looked up at me. I raised the touchdown salute. She said, 'Ron, do you have any news for us?'" Ron was handed a microphone and announced the approval of omaveloxolone for FA—the first disease associated with mitochondrial dysfunctions to have an approved treatment. There was a standing ovation, there were tears and hugs.

In Michigan, Ruth Acton was walking her dog when she got the text message about the approval. "I cried," she told us. Then she rushed back to share the news with Jack, and they celebrated with a bottle of champagne that was left over from New Year's. To some, the improvements brought by Omav may seem minor, but for someone like Jack, being able to continue to hold his own glass and bring it up to his mouth by himself is huge.[2]

A key point is that the knowledge gained from this rare mitochondrial disease can ultimately be used to treat common diseases or even aging; decline in NRF2 seen in FA patients and the associated increase in oxidative stress also occurs as we age.[3] This doesn't mean that Omav, specifically, can be used to treat aging, but that it can open the way for similar drugs. As we mentioned in chapter 3, a growing body of evidence associates age-related diseases with the quality of the mitochondria. These are diseases that become more common as people get older: diabetes, cancer, and cardiovascular and neurodegenerative diseases. Interventions that will improve the health of our mitochondria can delay the onset of these diseases and decrease their severity.

A Possible Shift in Medicine

Omav is possibly the bellwether for something much bigger. In the old days, before the surge of research in the field, and when not much was known about the role mitochondria play as mediators of diseases, scientists didn't see treating mitochondria as a possible solution. As researchers identify which pathways in the cell are not working in certain diseases, they develop a better understanding of what needs to be fixed and what specific molecule in the cell they should target. For example, the understanding of how frataxin dysfunction leads to oxidative stress in FA and the role of NRF2 in reducing oxidative stress paved the way for Omav.

There has been remarkable progress in the past two decades in our basic understanding of mitochondrial biology. Based on the new knowledge, scientists are exploring how to treat diseases that are related to mitochondrial dysfunctions, and in the coming years, we may experience a significant shift in medicine that aims to improve the functioning of our mitochondria, and thus our health. We're talking about a shift from organ-oriented medicine to a more holistic approach.

Medicine as we know it today emerged around anatomy, so we have specialists for every part of the body. Gastroenterologists for our gut, nephrologists for our kidneys, dermatologists for our skin, neurologists for our brain, cardiologists for the heart, and so on. The knowledge of these specialists is, and will continue to be, invaluable. But in many cases, although disease symptoms may be apparent in one organ, the problem may be in all organs, and by fixing mitochondrial functions in the diseased organ (or anywhere else), the active mitochondria will help treat the disease. And remember, when mitochondria are there to help, they not only provide the energy the cell needs to repair malfunctions, but they work on multiple fronts: providing building blocks, signaling, killing cells that should die, protecting from viruses, detoxifying pollutants, managing the yin-yang of oxidants and antioxidants, and more.

Doug Wallace, who started studying the link between mtDNA and disease more than fifty years ago, sees many common diseases as systemic disorders of energy and metabolism. And he believes that understanding mitochondria holds the key to their treatment. "If we just focus on an individual organ, we aren't able to effectively understand these diseases and therefore to diagnose and treat them properly," he told us. Currently, most medications treat *symptoms* and only very few drugs cure a disease, but with the new knowledge about mitochondria, we may finally be able to help the cell with functioning mitochondria that can treat the root problem in certain ailments and help the cell heal itself in others. We are only as healthy as our mitochondria, and having functional mitochondria is the key to our well-being.

There are other examples demonstrating that targeting mitochondrial functions can help address the medical problems in our body, irrespective of the cause of the specific disease. The first case that illustrates this point is metformin, a drug that is being widely used for type 2 diabetes. Researchers in the U.K. compared seventy-eight thousand people with early diabetes who had just been put on metformin with a control group of seventy-eight thousand people who didn't have diabetes or other diseases at the onset of the study. Remarkably, in addition to protection of the diabetics on metformin from early death, the researchers found that people with prediabetes on metformin had a lower mortality rate than the nondiabetics who didn't get metformin.[4] This and other studies have shown that patients who take metformin are less likely to have cardiovascular diseases, cancer, and other common diseases. There are several possible mechanisms by which metformin benefits the mitochondria.[5] Some scientists believe that the drug triggers a cascade of events that ultimately promote autophagy and mitophagy. Other researchers suggest a direct effect on Complex I in the electron transport chain. Regardless of the exact mechanism, like any drug, metformin has side effects, so it should be taken only if prescribed.[6]

In chapter 3, we met Nir Barzilai, who talked about mitochondrial dysfunction as one of the hallmarks of aging, and about research that shows that improving any one of the hallmarks will frequently benefit others.[7] A few years ago, Barzilai and a group of other gerontologists started to wonder whether they could delay or even prevent the onset of all age-related diseases. "Treating one disease at a time or targeting just one organ rather than targeting aging is a miserable approach—and it's not working," Barzilai argues.[8] They tossed around many ideas and drugs and eventually selected metformin because it is cheap, with minimal side effects, and most important, as the study above suggests, it works. And so, the idea was born of TAME—the Targeting Aging with Metformin trial. There was just one small problem: aging is not recognized by the FDA as a disease, so how can they run a clinical trial that targets aging? In 2015, they met with FDA officials to get their blessing for the trial and got a nod of approval.[9] Ten years later, however, they are still looking for funding; they tried both public and private sources, to no avail. "I don't know if we'll succeed, but we'll die trying," Barzilai says. Time will tell what will happen with this trial, but in our context, the main point is this: Experiments done with metformin demonstrate that by treating mitochondria, it is possible to prevent or postpone age-related diseases such as cancer, neurodegenerative disease, and heart disease. These are very different diseases with very different biological pathways, and yet it seems that improving mitochondrial functions does something to help all these ailments.

Daria got to experience firsthand the beneficial effect of mitochondria on multiple problems during a project in her lab dealing with excessive mitochondrial fission. Recall that mitochondrial fission is normal; it is part of that fission/fusion dance that ensures mitochondria quality. However, in numerous diseases, mitochondria split into such small fragments that they stop functioning properly and instead cause further cell injury through ROS. The project was initiated in 2010 by a postdoctoral fellow named Xin Qi (now a professor at Case Western Reserve University) who used a peptide (a short protein) called P110, which had been developed in Daria's lab (disclosure: Daria holds patents related to P110). Xin found that P110 stopped excessive fission in mitochondria in neuronal cells cultured in dishes.[10] Xin and Daria then tested P110 in a mouse model of Huntington's (a debilitating and fatal neurodegenerative disease).[11] They treated some of the mice with P110 and some with a placebo, and the results were intriguing: As the disease progressed, mice with Huntington's could hardly move, whereas mice with the same disease who were treated with the P110 peptide ran around their cage almost like healthy mice. (Eventually, though, the P110-treated mice succumbed to the disease, too.) In subsequent experiments, they showed that P110 greatly reduced excessive mitochondrial fission in neurons derived from humans with Huntington's.

And here's the kicker: By treating the mitochondria, the same compound has helped with a range of other diseases, too. P110 was then tested in animal models and in cultured cells of patients with Alzheimer's, Parkinson's, and amyotrophic lateral sclerosis (ALS, also known as Lou Gehrig's disease, named after the famous baseball player).[12] Again, these are very different diseases with different symptoms and biological pathways. But in all these cases, inhibiting excessive mitochondrial fission somehow enabled the cell to fix itself (at least partially), regardless of the cause of the pathology and greatly delay the progression of these diseases in animal models. Based on the initial studies, other academic labs have independently confirmed the efficacy of P110 in Parkinson's disease, multiple sclerosis, stroke, inflammatory bowel disease, acute kidney injury, traumatic brain injury, and more.[13] Daria has been trying to bring P110 to patients for a decade now, but drug development is no walk in the park and the first attempt to start a clinical trial for P110 was short-lived. However, the dream of testing this peptide in patients hasn't died, and regardless of what will happen with this compound, our main point is that treating mitochondria benefits the body and has the potential to reduce the burden of many diseases.[14]

There's another fascinating question that this research raises: How is it that multiple organs are affected? The P110 peptide was injected under the skin of the mice, so how did it affect the brains, for example? The answer so far is that we don't know. Yet reports of the same phenomenon where mitochondrial intervention in one area of the body influences mitochondria in another organ keep coming. Andrew Dillin from UC Berkeley, for example, found in worms that when he and his team fixed a problem in the mitochondria of certain nervous system cells, the problem was also fixed in the worms' intestines and extended some of these worms' lives by 20 percent.[15] While the exact mechanism is still unclear, these studies are consistent with what we described in chapter 2—that mitochondria's influence is not limited to their home cell and that it extends to other cells, tissues, and organs. This is another advantage of tending to mitochondria. Next, we'll see how practitioners take advantage of this fascinating trait of the life machines.

Mitochondrial Transplantation: Sending Life Machines to Help Other Cells

The groundbreaking field of mitochondrial transplantation offers a glimpse into how mitochondrial research may transform the future of medicine. Consider the following case: Shortly after baby Avery was born in 2016, it became clear that her heart wasn't functioning correctly, and she underwent open-heart surgery at Boston Children's Hospital. Her surgery was successful, and she was sent home, but a couple of weeks later, her situation worsened, and Avery was rushed back to the hospital. A heart transplant seemed to be the only way to save her life, but realizing that her heart tissue could possibly be revived, her surgeon, Sitaram Emani, suggested to her family an experimental alternative. As we discussed, mitochondria don't always stay in their original cells but can move into other cells and affect them, and some scientists and practitioners try to emulate these natural processes. Since 2015, Emani has been collaborating with another researcher, James McCully at Harvard, who developed a way to isolate healthy mitochondria from one part of the body and transplant them directly to another part. Avery's family decided to try this transplantation procedure.[16]

The idea of mitochondrial transplantation is nothing short of mind-boggling; we've all gotten used to hearing about organ transplantation, but the idea that a tiny organelle can be transplanted and function normally in another cell is cutting-edge. So we were anxious to hear more about it, and we met James McCully for coffee to hear a little more about this procedure. A tall, friendly, gray-haired Canadian, McCully told us about the first time he saw the potential of this concept. Medical students he was training operated on a pig's heart (which is similar to the human heart), and that pig went into fibrillation (irregular heartbeat). McCully's lab had some mitochondria they had isolated from that pig, and he suggested they inject the isolated mitochondria into the heart. Once they did, things transformed pretty quickly. "The pig's heart went from sort of a brownish color into pink and started beating away, and it kept beating for a couple of hours," McCully said. When he first presented the idea at a cardiology meeting, he faced major skepticism. One of McCully's students presented a poster summarizing the concept and the initial findings. McCully remembers one prominent scientist glanced at the poster, uttered, "That'll never work," and walked away.

Yet the animal experiments told a different story, and McCully's work got the attention of Pedro Del Nido, the chairman of the Department of Cardiac Surgery at Boston Children's Hospital. Del Nido thought that this procedure may address a problem he and his colleagues were facing. The tiny hearts that surgeons like him try to fix don't always wean fast enough from the machines that replace the heart and the lung during the surgery. Known as extracorporeal membrane oxygenation (ECMO) machines, these instruments are of great help during surgery, but kids who don't wean from them within seventy-two hours have a low chance of recovering. Del Nido got McCully in touch with Sitaram Emani, and the two started working on trying this on humans.

During the first operations, McCully got a lot of exercise. To prepare mitochondria for transplantation, Emani would give him a small piece of muscle tissue taken from the patient. McCully would run down three flights of stairs, dash outside, and climb four flights of stairs to his lab, where he had the equipment to isolate mitochondria from the tissue. Then, he would put the freshly isolated mitochondria in a test tube with a bag of ice, run down four flights, and then up to the operating room on the third floor, where Emani would inject the healthy mitochondria into the patient's heart. Nowadays, McCully has a well-equipped bench in the operating room where he can isolate a patient's mitochondria more quickly, and the healthy mitochondria seem to integrate into the heart muscle cells and make a difference. As of May 2024, sixteen children have undergone this procedure, and the percentage of those kids who were able to come off ECMO machines has risen dramatically, compared with historical data.[17] Avery, who's turning nine as this book goes to print, is one of these kids. She's had additional surgeries since then and still needs monitoring, but the injection of mitochondria saved her heart, and her life.

Over at the University of Washington in Seattle, a neurological surgeon named Melanie Walker heard about McCully's work and had a hunch that mitochondrial transplantation may address a problem she and her colleagues face with stroke patients. In cases of ischemic strokes (when a blood clot blocks an artery that supplies blood to an area of the brain), the standard of care is to insert a catheter through the groin to pull out the clot. The act of restoring the blood flow to some tissue or organ is called reperfusion, but it can come with a heavy price for cells—*reperfusion injury*. We mentioned this term when we discussed heart attacks in the previous chapter: Paradoxically, a cell that was starved for oxygen can be damaged or die from a sudden flow of oxygen. Walker was intrigued by the fact that mitochondrial transfer is something that is already happening naturally in the body, and transplantation simply amplifies it; as mentioned in chapter 2, during a stroke, neighboring cells help damaged neurons by sending them fresh mitochondria. Walker started talking about the idea with her colleague Michael Levitt, who was initially very skeptical. Yet Levitt felt that testing the idea would be important because, like Walker, he was concerned that reperfusion injury is not being addressed by any treatment.

"Simply put, right now all we do is plumbing," Levitt told us. "By unplugging the clot, we turn on the sprinklers on the lawn, but we can't undo the damage to the grass. We can restore blood flow, but we can't repair on a cellular level any damage that has already been done." With the help of colleagues at the University of Virginia, Walker and Levitt ran some experiments in mice with very convincing results. Mitochondria that were injected into the stroke area in those mice integrated into neural and glial cells and increased ATP levels, reduced the volume of the dead or injured area of the brain, and increased neuronal viability.[18] Seeing those results was a big "aha" moment for Levitt, and he and Walker have started a clinical trial. During the procedure, they take a small bit of tissue from the patient's muscle and pass it on to a scientist who then isolates mitochondria from the harvested tissue. The healthy mitochondria are then injected into the brain through the catheter that is used to extract the clot. It remains to be seen if this application of the concept is successful, but Walker and Levitt say that the initial results are encouraging.

The field of mitochondrial transplantation is in its infancy, and it still faces skepticism. Indeed, many questions are left open. For example, researchers have doubted that mitochondria can survive in the blood, a high-calcium environment that they are injected into. Yet the concept continues to show good results in other fields, including cancer, organ transplantation, and in patients with primary mitochondrial diseases.[19]

Some of these studies suggest that mitochondrial transplantation may have positive effects across the body, not just on one organ at a time. An Israeli company, Minovia Therapeutics, demonstrated this in a small clinical trial with children with mitochondrial diseases. When the company's founder, Natalie Yivgi-Ohana, first described her concept, many dismissed it. She wanted to harvest stem cells from kids who suffer from serious mitochondrial diseases, augment them with purified mitochondria taken from their mothers, and then give those cells back to those kids. Yair Anikster was among the skeptics; Anikster runs the Pediatric Metabolic Disease Unit at the Sheba Safra Children's Hospital in Israel, and when Yivgi-Ohana first told him about the idea, he didn't buy it. "But after I saw the first patient who was treated, I started to push them to treat more patients," he told us. This was a seven-year-old boy with Pearson syndrome (a rare mitochondrial disease); he was anemic, didn't gain weight, and had kidney disease and cognitive problems. After the treatment, the boy showed significant improvements on all these fronts. He gained weight, his cognitive performance improved, and he had more energy. In the small clinical trial with six patients, the researchers found that the kids who were treated benefited in several ways, including increased mitochondrial quality, improvement in motor function, weight gain, and better muscle and cardiac endurance.[20]

The current procedures of harvesting mitochondria are highly specialized and can be executed only on a small scale, so several companies around the world (including Minovia) are trying to create stocks of isolated and stored mitochondria from different healthy people that will allow much wider use for the medical conditions mentioned above. In a 2024 study, Jonathan Brestoff from Washington University in St. Louis and Japanese collaborators (including scientists from LUCA Science, a Japanese company) have done just that. The researchers showed that a weekly injection of stored frozen human mitochondria into the bloodstreams of mice with a severe mitochondrial disease due to a large deletion in mtDNA (representing Leigh syndrome in humans) greatly prolongs the lifespan of these sick mice. It also improved the function of many organs, from lungs and muscles to heart and brain. So stored mitochondria from both healthy mice or healthy humans were beneficial in sick mice and triggered no rejection.[21] These data are exciting, since they suggest that it is not necessary to transfer only freshly isolated mitochondria from the same patient that is their recipient; the study suggests that it is possible to generate a "universal donor pool" of mitochondria, stored for future use for any patient in need.

While encouraging, there was something in this study that highlighted that there are still many open questions in the area of mitochondrial transplantation. Although these mice had a severe central nervous system dysfunction, which was improved by injecting healthy mitochondria in other parts of the body, the researchers found no evidence for these mitochondria in the brains of the treated mice. We faced a similar conundrum earlier when we described Daria's research on P110. And this, too, is the type of finding that makes researchers scratch their heads: How can the mitochondrial transfer show benefit to the brain, and yet no transferred mitochondria were found in the brain? Brestoff and his colleagues suggest three potential explanations: First, the effect may be due to the entry of mitochondria into peripheral neurons (those outside the brain). Second, it may be that the donated mitochondria increase the level of antioxidant enzymes, thus improving overall neutralization of ROS in the body. Third, it is possible that transplanted mitochondria trigger mitophagy of damaged mitochondria.

Can mitochondria transfer one day apply to dealing with aging? Will we ever see bags of high-quality mitochondria in hospitals ready for transfusion? Some scientists believe that it's possible, while others see it as science fiction.

Regardless, the early successes in transplantation illustrate an important point: The potential of a relatively small number of healthy mitochondria to heal, by tipping the balance in favor of healthy mitochondria. The pioneers in the field of transplantation achieve this by placing pristine mitochondria in the body, but it can be done in other ways. For example, in chapter 6 we'll see that exercise increases the number of well-functioning mitochondria in skeletal muscles. This means that a good balance between healthy and damaged mitochondria in the cell can also be achieved through lifestyle changes.

mtDNA Manipulation:
Tipping the Balance to Help the Cell

The idea that you don't have to eliminate every single bad mitochondrion to help the cell is critically important. And this concept has implications both for medical intervention and our personal choices in life. Is there anything that we can do ourselves to remove bad mitochondria? Yes, and this is what part II of the book is all about. There, we won't talk about medical procedures but about lifestyle changes. As we pointed out with the example of exercise, sometimes a small change in our lifestyle can tip the balance in favor of the good mitochondria in our cell, thus improving its health. If this happens in enough cells, it improves the health of the whole body.

Before we move on to part II of this book, let's first look at one more example of how scientists hope to take advantage of this concept in developing medical interventions. Since there have been great advancements in editing nuclear DNA in the past couple of decades through tools such as CRISPR, scientists hoped that they could use similar approaches to edit mtDNA. You can think of CRISPR as a pair of scissors that can cut nuclear DNA at a precise spot and then stitch in a "fixed" part, a piece of DNA that lacks the mutation. CRISPR has great potential for treating many disorders, including some related to mitochondria. In fact, in 2025, an infant who suffered from a rare life-threatening mutation in a mitochondrial enzyme involved in the urea cycle was reported to be the world's first patient to undergo personalized gene editing using customized CRISPR.[22] In that case, CRISPR was used to edit nuclear DNA. Can this method be used to edit mitochondrial DNA?

Unfortunately, so far this approach hasn't worked with mtDNA, for a variety of reasons. Yet somewhere along the line, scientists realized that maybe they don't need to edit the mtDNA; they could instead manipulate it differently by taking advantage of a unique property of mitochondria—their number. Each cell has hundreds or even thousands of mitochondria, and each mitochondrion contains several copies of mtDNA that are continuously replicated.

Maybe, scientists started to wonder, what they need to do is just reduce the number of the mutant mtDNA. A scientist in this field, Carlos Moraes from the University of Miami, who's been using this approach, explained: "You really don't have to eliminate mutant mitochondrial DNA completely." You just need to move it below the pathological threshold.[23] As long as the number of mitochondria with mutated DNA stays under a certain threshold, a disease will not manifest itself. Moraes and other scientists in the field, like Michal Minczuk at Cambridge, have developed different methods to selectively remove the mutant mtDNAs, and have shown that after they do it, the cell repopulates itself with mostly healthy mitochondria.

Can mtDNA manipulation apply to age-related diseases? The current focus is on addressing specific mutations in mitochondrial diseases, and more research is needed to understand whether age-related diseases are derived from such mutations.

● ■ ●

This chapter (as the rest of the book) is a snapshot of current research, and things may evolve in ways that are hard for us to imagine today. New discoveries may lead to new drugs, mitochondrial transplantation may be enabled by some breakthrough technology, or a totally new way to manipulate mtDNA may emerge. We can't predict the future, but the bedrock of our belief is that an organelle that plays such a central role and has so many functions in the cell cannot be ignored when it comes to healing the body. So focusing on mitochondria can have a big impact on human health.

And this takes us to part II of the book. We saw how scientists seek to help the cell through medical interventions in three ways: by reducing the burden on the cell, by taking advantage of the fact that mitochondria can help other cells, and by keeping damaged mitochondria under the threshold. But wouldn't it be even better not to reach the point where we need these medical interventions to begin with?

We can benefit greatly from tending to our mitochondria through our lifestyle choices. When we exercise, eat well, sleep well, reduce prolonged stress, and avoid toxins, we improve our mitochondrial functions and thus reduce the burden on the cells. We also increase the chance that the healthy mitochondria in one part of our body will help neighboring cells or even cells in other parts of the body. And we can possibly reduce the number of damaged mitochondria in each cell and thus bring the number of "bad" mitochondria under the threshold. We're all born with our genetic makeup, and we can't change that, but research shows that our health can be greatly determined by our life choices.[24] What we do and how we live our lives can either help our mitochondria or hurt them, and by helping our mitochondria, we may be able to help our body even before symptoms show up. To paraphrase JFK, our focus in the second part of the book is the following: Ask not what your mitochondria can do for you—ask what you can do for your mitochondria.

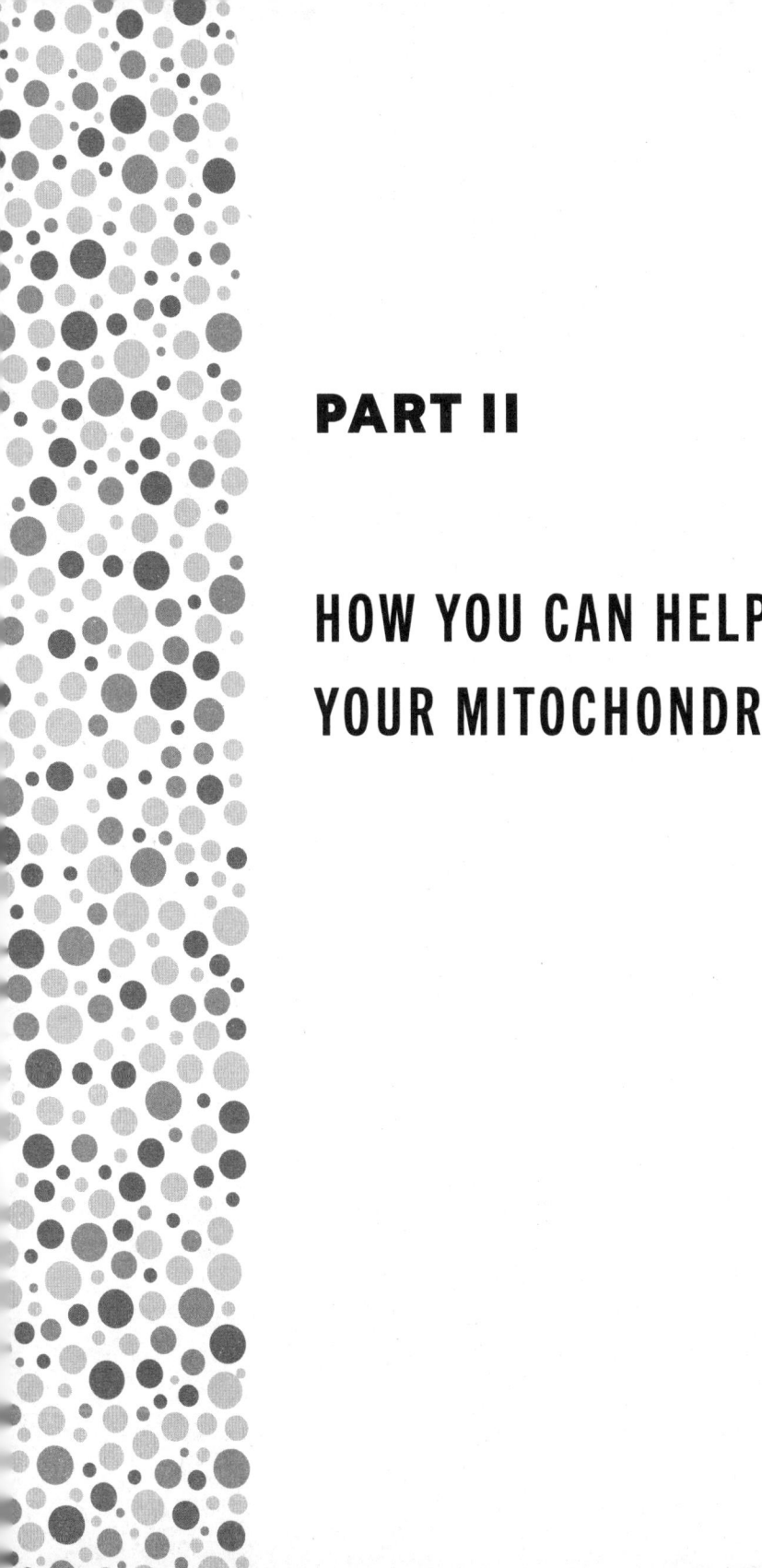

PART II

HOW YOU CAN HELP
YOUR MITOCHONDRIA

Now, let's see how you can help your life machines help you. In this part of the book, we'll describe some lifestyle changes that you can make to be kind to your mitochondria. As we transition to this section, we'd like to point out seven guiding principles that go beyond any specific change.

1. **There's more than one way to live a healthy life.** Your mitochondria will benefit from your exercising, eating healthy food, getting a good night's sleep, and reducing stress. But there isn't a single diet or one exercise regimen that will get you there.

2. **The Goldilocks principle.** Research consistently shows that avoiding excess or deficiency in biology is the way to optimal health. This principle is named after the little girl who liked her porridge at just the right temperature—neither too cold nor too hot—in the famous fairy tale "Goldilocks and the Three Bears."

3. **The answer may be "both."** People can get into lengthy arguments that include the word *or*: Which is more important: endurance exercise or resistance exercise? Carbs or proteins? In many cases, the answer is "Both."

4. **There's no magic pill.** The idea that one drug, one method, or one of anything will be helpful for all is extremely unlikely to be correct.

5. **A lot is still unknown.** There's still a lot that we don't know about mitochondria (and life in general). This book is a snapshot of the current knowledge, and there is much more that remains to be discovered.

6. **We don't have full control.** As humans, our brain seeks simple explanations, so we'd like to think that if we do X, we'll get Y. Because of this addiction to simple answers, some people who get sick immediately blame themselves ("I'm sick because I didn't do X"). Yet our body is immensely complex, and the outcome of our actions is also affected by our genetic makeup, our past, what we have been exposed to, what access to health care we have, and, of course, chance.

7. **It's never too late (and it pays not to wait).** While we don't have full control over the outcome, we can increase the likelihood of positive outcomes through what we do and what we avoid. It's never too late to do things that may help your mitochondria help you. Research also shows that starting early pays off.

CHAPTER 6

Go, Kyle, Go!
Mitochondria and Exercise

When we told people that we were writing this book, they often asked how they could take care of their own mitochondria. There are several habits you can adopt that may help you achieve that, and we'll explore them in the pages that follow. Regular exercise is probably the one habit that is backed by the strongest evidence. Simply put, when you engage in exercise, there is a higher demand for energy, so your body increases the number of well-functioning mitochondria in your skeletal muscles in preparation for the next time. This was shown in the 1960s by a young researcher named John Holloszy.[1] Early on in his career at the University of Illinois, Holloszy helped with a study of a group of middle-aged men who were part of an exercise program, and he was impressed by the progress they were making. When the program started, participants could run a quarter of a mile, and after two or three months of training, they were running four miles at a pretty good pace. Researchers believed that the improvement in exercise performance was due entirely to an increase in the heart's output, which now provided the skeletal muscles with more blood and oxygen. But Holloszy thought that this alone could not explain the improvement. Looking for answers, he came across a journal article that had compared muscles from wild rabbits to domestic rabbits, and another paper that compared muscle from wild ducks to domestic ducks. Both papers pointed out a huge difference in mitochondrial content of the wild and domestic animals, which Holloszy suspected was due to the difference in their physical activity. After all, wild ducks fly and have to fight for their survival, whereas domestic ducks are bred to have large bodies and small wings. Wild rabbits roam the prairies, whereas lab and pet rabbits are confined to a cage. Holloszy became fascinated by the idea that physical activity may affect mitochondria in muscle and looked for ways to prove it.

When he set up his lab a few years later at Washington University School of Medicine in St. Louis, Holloszy and an assistant started training rats to run on a treadmill. When they dissected the rats' muscles after three months of training, they noticed that the muscles of trained animals had a much deeper red color than that of the untrained animals. Further analysis showed that there was about a twofold increase in mitochondrial proteins, indicating an increase in functional mitochondria.[2] Follow-up studies have confirmed that endurance exercise improves skeletal muscles by increasing the number and quality of mitochondria in these muscles. Exercise has another benefit that is especially important as we age: It stimulates autophagy and mitophagy, which as we explained earlier, are the processes by which damaged and fragmented cell components are isolated and recycled into new cellular structures.[3] It also leads to an increase in antioxidant protection, which means that our cells are better prepared to handle free radicals.[4] Bottom line: Exercise improves mitochondria in skeletal muscles. But the benefits of exercise don't stop there.

Effect on the Heart, the Lungs, and the Brain

We must make a little confession here. By talking so much about the value of exercise, we may create the impression that we're both fitness nuts who start the day at the gym, hop into the pool, and continue with a 5K run. We are not. We do exercise regularly, but we are not enthusiasts, and we're very familiar with the internal dialogue that may go on in your mind as you're trying to motivate yourself to get off the chair. So we approached the research on this with trepidation. Who are we to talk about the value of exercise? Going over dozens of papers on this topic took us much longer than on other topics, possibly because they kept reminding us that we were not doing enough. For a little help, we reached out to Nadège Zanou from the University of Lausanne, Switzerland, who had coauthored a review article on the role of mitochondria in exercise-induced neuroprotection. Zanou is an MD/PhD who also conducts research in the field, and she patiently introduced us to the exciting world of myokines.[5]

Myokines are signaling molecules that are released by muscle fibers when you exercise. Some researchers call them exerkines to emphasize that they are released after exercise, and a growing body of evidence suggests that they influence several organs, including the heart, the liver, fat tissue, the lungs, and, most important, the brain. Zanou started by talking about brain-derived neurotrophic factor (BDNF) and its effects on the development of new neurons in the brain, learning, and memory formation, which are well established. Research indicates that many of the myokines that muscles release during exercise affect the expression of BDNF. Covering all the myokines is beyond the scope of this book, so instead we'll focus on one—irisin.

It all starts with a protein called PGC-1α, the master regulator of the process that increases the number of mitochondria.[6] A group led by Bruce Spiegelman from Harvard Medical School reported that exercise increases the expression of this protein, which eventually leads to the release of irisin. The name wasn't chosen at random. Iris was the Greek messenger goddess, and the researchers thought the name reflected its role—sending messages from the muscles to other parts of the body. The group found initially that aerobic exercise stimulates this signaling protein to start a cascade of events that converts white fat tissue into brown fat, which produces heat when you're cold and reduces the risk of obesity. (Brown fat gets its color from abundance of mitochondria.) Spiegelman's group also observed that a ten-week program of endurance exercise significantly increased the level of irisin in the blood of humans.[7] Follow-up studies reported the benefits of irisin in the heart, the lungs, and the liver.[8] In 2021, Christiane Wrann from Massachusetts General Hospital and Harvard Medical School reported some interesting results together with Bruce Spiegelman.[9] The researchers manipulated the genome of mice so that those mice were unable to produce irisin. We'll call these mice the "no-irisin mice." Wrann, Spiegelman, and their colleagues let mice run on wheels and found that exercise did not improve the cognitive ability of the no-irisin mice, as opposed to normal mice, whose memory and learning improved. Also, as opposed to the normal mice, the no-irisin mice developed more cognitive problems as they aged. Now comes the part that we find most interesting. Wrann and her colleagues developed a way to artificially increase the level of irisin in the no-irisin mice that had already developed significant pathologies as they aged. After two months they found that no-irisin mice that went through this treatment experienced a significant improvement in their cognitive functions.

As with other new concepts that we have described throughout the book, there are still open questions and heated debates about irisin. (For example, what is true in mice may not necessarily be true in humans.)[10] However, from a practical perspective, any such debate should not discourage you from exercising because there is very strong evidence associating exercise with positive effects on the brain. For example, exercise improves blood flow to the brain,[11] and studies report a strong association between exercise and lower risk of dementia even in older adults.[12] If you're still not convinced about the value of exercise to your brain, studies suggest that exercise moderately increases the level of humanin in our body, and this mitochondrial-derived peptide has been shown to have neuroprotective properties. The bottom line is this: We can all protect our brain through exercise, and starting even later in life can be beneficial, as long as we stick to it.

Pushing the Limits

It's customary to lament about Americans who don't exercise enough. Indeed, only about a quarter of us meet the recommendations of two weekly sessions of some muscle-strengthening activities (such as weight lifting) plus 150 minutes of moderate-intensity aerobic physical activity or 75 minutes of vigorous aerobic physical activity.[13] Yet when you think about it, it's a miracle that even a quarter of us meet the minimum exercise recommendation because we evolved to conserve energy, not to waste precious ATP on treadmills. As Daniel Lieberman from Harvard explains, for most of human history, people had to work hard to find food or run away from predators, so conserving energy by resting makes perfect sense.[14] "Our minds never evolved to get us moving unless it is necessary, pleasurable, or otherwise rewarding," Lieberman wrote. Think about it next time you resist getting up from the sofa. It's not your fault, and later we'll talk about what can be done about it.

Given this evolutionary tendency, we're in awe of people who run marathons, do triathlons, swim across the English Channel or cross anything else, for that matter. Which brings us to the story of a man named Kyle Bryant and his dream to cross the United States on his red tricycle. As a teenager, Kyle started to frequently lose his balance. After numerous stitches from stumbling or falling over on his bike, he underwent neurological examinations and eventually was diagnosed with Friedreich's ataxia (FA; this is the disease we discussed in the last chapter; it stems from the lack of frataxin in the mitochondria). After graduating with a degree in civil engineering, Kyle got a job with an engineering firm in Sacramento, California, but he felt that something was missing.[15] Then, one day, he tried a recumbent trike, and the moment he put his feet on the pedals he felt in control again. He could ride without wobbling and stop without falling over. After just a few weeks, he took his first fourteen-mile ride, and as he kept increasing his distances, he developed a plan to ride twenty-five hundred miles across the country. In 2006, Kyle attended a meeting of ataxia patients in Boston, and he told people about his wish to ride as a way to raise awareness of Friedreich's ataxia.[16] A father of another kid with FA offered to give him one dollar per mile, and other families chimed in.

On January 22, 2007, Kyle started the ride from La Jolla, California, to Memphis, Tennessee, where he was supposed to give a talk at the National Ataxia Foundation annual conference. With a helmet on his head and colorful wristbands from supporting organizations on his wrist, Kyle sat on his trike and started the ride. Soon after leaving the San Diego area, he headed toward the mountains, which demanded a climb from a few hundred feet to four thousand feet. During training, a tendon in Kyle's leg became inflamed, and after the grueling climb the pain in his knee intensified, making each pedal rotation harder and harder. Still, he was inspired, and while he was exhausted, he found the time at the end of each day to blog for his supporters. Navigating between road kill as eighteen-wheelers passed him at sixty-five miles an hour was scary at times, but Kyle felt that he was making a difference, raising attention and money for a rare disease. Yet after a few days, the immense physical effort started to take its toll. At some point, his fatigue and the knee injury intensified, and he was again engulfed in self-doubt. "Who am I kidding? This is not something I was meant to accomplish."

We'll come back to Kyle later in this chapter, but this is a good point to talk about personal fitness. When it comes to exercise (as with everything else), each of us has a different starting point that is determined by our genetic makeup and by our past. Some are Olympic athletes, while others can hardly move because of a severe genetic or chronic disease. Not everything is under our control, and we should consider this when thinking about engaging in physical activity.

Which Type of Exercise Is Best for Your Mitochondria?

From the initial research that found that wild ducks have more mitochondria in their muscles to recent studies that demonstrate the positive impact of exercise on our brain, research indicates that exercise helps the life machines. The obvious question that follows is: What type of exercise is best for me? The answer can vary based on your health, age, previous experience, and— importantly—what brings you joy. (Recall that we're programmed to conserve energy and spend it when it is necessary, but also if it is pleasurable or otherwise rewarding.) You should talk to your primary health-care provider before you adopt a new exercise routine, and before adopting a regimen that involves high-intensity aerobic exercise, your health provider may recommend a cardio stress test. Also, as important as mitochondria are to your health, their condition is just one of several considerations in choosing an exercise program, so ultimately, you and your physician will decide what's best for you.[17]

When it comes to brain health, studies in mice and humans suggest that most benefits come from endurance (aerobic) exercise—activities that increase your heart rate and breathing such as jogging, climbing stairs, brisk walking, biking, or swimming. In the irisin experiments mentioned above, the mice didn't go for leisurely walks or lift weights. They ran. Other types of exercise have their benefits, too (for example, resistance exercise is critically important to curb muscle loss), but endurance exercise seems to be the best way to help your brain. This doesn't mean that the only way to get there is on a treadmill or a stationary bike. Researchers often use these machines because it gives them more control over different variables, but your mitochondria don't care if you get your aerobic exercise by dancing, hiking uphill, taking a brisk walk, or swimming.[18] Physiologists and coaches use a scale of "zones" to describe this, and while the definitions that are used by different experts are not always identical, the general idea is that Zone 1 is the easiest and Zone 5 is the hardest.

One way to assess which zone you're in is to compare it to your maximum heart rate, which is roughly computed by subtracting your age from 220. (So for example, if you're fifty, your maximum heart rate is 170. Certain online calculators use different formulas.) Some specialists describe the zones as they relate to your ability to talk. In Zone 1 (50 to 60 percent of your maximum heart rate), you can talk easily. In Zone 2 (60 to 70 percent), you can still talk, but you may need to stop from time to time to catch your breath. In Zone 3 (70 to 80 percent), talking takes effort. In Zone 4 (80 to 90 percent), you don't want to talk, and in Zone 5 (90 to 100 percent), you can't talk even if you want to.[19]

Using your maximum heart rate as a reference point is just one way to assess your effort. Another method is VO_2 max, which is considered one of the best ways to measure overall cardiovascular respiratory fitness. It is routinely used by top athletes and is gaining popularity in wider circles. It can be tested in sports medicine facilities, medical labs, and some gyms. Several wearable devices and smartwatches provide an estimate for it. How does VO_2 max relate to your mitochondria? It tells you the maximum capacity of your body to utilize oxygen, and, to some extent, it can serve as a good proxy for your mitochondrial function because to produce ATP, your body needs oxygen, so the rate of oxygen consumption can tell you about your ATP production. (It's a proxy and not a perfect measure of your mitochondria because VO_2 max can also be affected by things like the oxygen-carrying capacity of your blood, or your cardiac output.)[20] If you get a high score on your VO_2 max test, it implies that your mitochondria are producing lots of ATP, which is obviously a good thing. Your VO_2 max is age and gender-specific, and tables with some values can be found online.[21]

Can VO_2 max be improved? Studies consistently show that it can be done. For example, in a study conducted at Massachusetts General Hospital, twenty-four sedentary adults completed a twelve-week aerobic exercise program, and they significantly increased their VO_2 max and their cognitive capacity. The mean age of this group of mostly women was sixty, and to be clear, they didn't become Olympian athletes after three months of training. But their capacity to utilize oxygen improved, which suggests that their mitochondrial ATP production improved, too.[22]

There's a lot of talk about the benefits of high-intensity interval training (HIIT). Is it good for your mitochondria? There are different regimens of HIIT, but in general, they include several minutes of high-intensity movements that increase the heart rate to at least 80 percent of one's maximum heart rate. These bursts are followed by short periods of lower-intensity movements. One of several studies aimed to assess which type of exercise is better for your mitochondrial functions and muscle strength and mass was conducted at the Mayo Clinic.[23] In that study, sixty individuals were assigned to three different exercise programs that each lasted twelve weeks: Participants in the first group did high-intensity interval training that included cycling. Those in the second group did weight training, and those in the third group did both cycling and weight training. Importantly, the latter group did the cycling and weight training at a lower intensity than the other two groups. The researchers then took biopsies from the participants' thigh muscles. Not surprisingly, the participants who did weight training (the second and the third groups), increased their muscle mass and strength. However, when it came to mitochondrial functions, the high-intensity interval training group had the highest gains. Does this mean that HIIT is the only way to improve mitochondrial functions? No, because those who did both cycling training and weight training had some nice gains, too. This and other studies suggest that HIIT is not the only way to go and that since maintaining muscle mass is important, a mixed regimen may be the best way.[24]

Exercise seems to also stimulate the secretion of certain mitochondrial-derived peptides ("little treats"), and here, too, scientists are trying to determine which type of exercise would do the best job. You may remember humanin, which has been shown to have neuroprotective properties. Scientists from Sweden recruited twenty participants, ten of whom did resistance exercises, and ten others did endurance excercises. The researchers found that the levels of humanin in the blood were significantly elevated by endurance exercise but not by resistance exercise.[25] Another beneficial mitochondrial-derived peptide, MOTS-c (which we met in chapter 2) was found in skeletal muscles and blood of humans four hours after some intense exercise (ten rounds of one-minute riding a stationary bike at maximal capacity with an interval of a minute and a half of rest). Recall that among its benefits this peptide has been reported to reduce obesity and extend health span.[26]

While aerobic exercise is important, that doesn't mean in any way that resistance exercise is not. Starting at around the age of thirty, we lose on average 3 to 8 percent of our muscle mass every ten years, and the rate of muscle loss (known as sarcopenia) is accelerated around the age of sixty.[27] Resistance exercise can curb this loss, and so far, physical activity remains the only proven way to deal with sarcopenia.[28]

The main takeaway so far is that exercise increases the number of well-functioning mitochondria in skeletal muscles, and that endurance exercise seems to benefit our brain. Resistance exercise is essential, too, in order to curb muscle loss. Adopting a regimen that includes both will increase the likelihood of extending health span, spending more of your older years as an active person rather than in bed or at the doctor's office.

What Doesn't Kill You Makes You Stronger?

Is there such a thing as exercising too much? Somewhere near the Texas border things got worse for Kyle Bryant. Facing a gradual uphill climb, one of those that seems never to end, he could no longer move his leg. His knee was out of commission. His father and his uncle pushed the trike uphill and loaded it onto the family's van. Kyle remembers it as a very scary moment. "My dad had to carry me to the bathroom because my knee was in such bad shape," he told us; "there was no way we were going to continue." So this was it. Reaching Memphis on a bike now looked like a dream that would never materialize.

Anyone who works out regularly knows about little (or big) injuries like pulled muscles or sprained ankles, and Kyle's injury can be seen as such. But some of the things that happen at the cellular level of all these injuries can involve our mitochondria, and scientists are starting to learn about the risk of pushing one's mitochondria too far. For example, in a small study conducted in Sweden, the researchers showed that excessive training caused mitochondrial impairment in healthy volunteers who went through four weeks of high-intensity interval training. Participants did fine after the first week of HIIT, and when the workload was ramped up in the second week, tests showed an improvement in both the volume and efficiency of their mitochondria. However, their mitochondria suffered on the third week when the weekly exercise was further raised to more than two and a half hours of HIIT. Their mitochondria showed some recovery after the fourth week of less demanding exercise, but not to the level of the second week.[29] Some experts in the field are not concerned about this problem, and although more experiments determining the specific effects of demanding exercise on the mitochondria are needed, other studies have already confirmed the risks in pushing too far. For example, in a 2003 study conducted in Israel, a group of thirty-one young and healthy men participated in a couple of grueling marches of thirty to fifty miles, carrying seventy-five pounds of extra weight. Blood samples were drawn before and after each march, and the level in the blood of certain chemicals associated with kidney and liver injury went through the roof.[30] The researchers pointed out that the utilization of oxygen in the muscle during strenuous exercise can increase significantly, and this may lead to much more electron leakage in the mitochondria and, as a result, more ROS. Recall that major damage and inflammation can occur when we produce more ROS than the antioxidant system can scavenge.

There have been other studies that demonstrate the risks in pushing too far, including increased acute inflammation. For example, a 2021 study showed that frequent strenuous exercise in people with a genetic risk for ALS (Lou Gehrig's disease), a disease associated with mitochondrial damage in motor neurons and neuroinflammation, increases the probability that they will develop ALS by sixfold.[31] Also, physicians who treat patients with primary mitochondrial diseases are well aware of the need to proceed cautiously. Exercise is usually good for their patients, but pushing too hard may lead to an exhausted patient who can't get out of bed for days. Of course, for most Americans, the problem is at the other end of the spectrum—lack of sufficient physical activity—but it's important to remember the Goldilocks principle here, too. No doubt that exercise is beneficial, but some believe that going too far can hurt you.

Things get even more interesting because it turns out that a little bit of damage in the cell can be beneficial; it activates a mechanism of repair that might help the cell better handle future damage. Known as hormesis, this idea is often associated with the saying "what doesn't kill you, makes you stronger," which is attributed to Friedrich Nietzsche. This is another effect that some scientists believe happens during exercise. Throughout the book, we've been emphasizing the damage that ROS can cause (and we just saw yet another example in the Israeli study above), but it turns out that a little bit of ROS caused by exercise might have a beneficial effect, by improving our antioxidant machinery.

To understand the basic science behind this phenomenon, consider the following experiment, which demonstrates hormesis in mitochondria, known as mitohormesis—the idea that short-term stress on mitochondria may lead to long-term adaptations. Gerald Shadel, a professor from Yale University, focused in this experiment on one type of ROS, called superoxide, which is a by-product of the energy transfer process. The antioxidant that gets rid of superoxide is an enzyme called superoxide dismutase (SOD), and Shadel and his colleagues developed a way to turn SOD on or off in mice. During their experiment, the researchers stressed some mice when they were still in their mother's uterus by turning SOD off briefly, which meant that they were exposed to more ROS for a while. For the control group of mice, the researchers did not turn off SOD. After the mice were born, the two groups of mice were similar in many ways, except for one striking thing: The mice that were briefly exposed to stress by turning off their SOD were in better shape than the mice who had not experienced stress. They had higher levels of antioxidants, more mitochondria, and less superoxide buildup.[32] Shadel didn't study this to show the impact of exercise but to demonstrate the basic concept of mitohormesis. Of course, it's a challenge to prove the link between exercise and mitohormesis in humans in such an elegant way. Yet when we met Shadel in La Jolla, California (he's now at the Salk Institute), and asked him what he does to protect his mitochondria, he didn't think twice: exercise.

There are studies in humans, too, that show a link between exercise and mitohormesis, suggesting that a little stress makes your mitochondria stronger.[33] For example, a study by researchers from Catholic University of Louvain, Belgium, showed evidence of damage to the muscle after two and a half hours of exercise, but this type of damage was lower among those who practiced endurance training long term.[34] Another study showed that exercise-induced stress generated more blood vessels over time and therefore improved performance.[35] In addition, a study from York University in Toronto showed that repetitive bouts of oxidative stress induced by exercise triggered increased expression of certain antioxidants, resulting in better capacity to neutralize them.[36] The bottom line: Pushing yourself (but not too much) can have beneficial effects.

It's Never Too Late (and It Pays Not to Wait)

Exercise is important at any age, and it is especially important to help our mitochondria as they deteriorate in our later years. Is it possible to have good mitochondria when we are older? There is overwhelming evidence that the answer is yes. For example, in 2019, researchers from Belgium recruited thirty-three men who were classified into four groups: young sedentary, young active, old sedentary, and old active. The sedentary men (whether young or old) had not been engaged in any weekly physical activity session for at least five years before the research started. In contrast, the active participants (again, whether young or old) were all experienced cyclists who reported at least six hours of training a week. The average age of all the young participants was twenty-two and the average age of all the old participants was sixty-seven. Analyses of skeletal muscle biopsies of both active groups showed that they had healthier mitochondria than the sedentary groups of similar age.[37]

Many other studies confirm that you can benefit greatly from exercising at any age, and we often hear that it's never too late to start, but is it true as far as your mitochondria are concerned? In 2018, researchers led by Francesca Amati from the University of Lausanne, Switzerland, put twenty-two healthy sedentary men and women (average age sixty-six) on an exercise program that included mainly biking and walking or running on a treadmill three times a week for four months. The researchers also recruited eight men and women around the same age who were lifelong exercisers. The previously sedentary participants enjoyed several improvements: the mitochondria in their skeletal muscles were younger than before because of their faster turnover, and their mitochondrial content even increased to the level of that of the lifelong exercisers. But not everything was the same. For example, while decreased fission, increased fusion, and increased mitophagy were recorded in this group, these processes were not at the same level as in the lifelong exercisers.[38] If you have lived a sedentary lifestyle, this finding should not discourage you because it suggests that your mitochondria will benefit from your starting to exercise; it's just that they won't necessarily improve to the same level as someone who's been exercising all their life.

Resistance exercise becomes especially important as we age. You may remember Sophie and Emma from chapter 3. Sophie was an active seventy-year-old while Emma was much less active, and her condition was defined as "pre-frail," which was reflected in the health of her mitochondria (Sophie's were in much better shape than Emma's). Muscle loss (sarcopenia) in the elderly is a serious problem, and as we pointed out earlier, it starts as early as at the age of thirty and accelerates from around the age of sixty.[39] Resistance exercise is widely used as an effective way for both strength and muscle mass enhancement in older individuals.[40] To be clear, you don't need to become a bodybuilder. In fact, research suggests that high repetition (with less weight), is the most effective strategy to protect against the problems associated with muscle aging.[41]

There's a critically important point for younger readers here. While studies show that it is never too late to start, it shouldn't lead you to the conclusion that you can postpone getting serious about exercise. It's a good idea to start exercising as early in life as you can for two reasons. First, as we've shown throughout the book, the health of your mitochondria is important at any age, so if you're in your thirties and plan to start exercising when you reach sixty, you run the risk of having suboptimal mitochondria for three decades. Not a good idea. The second reason is that, as we just saw, lifelong exercisers may have some advantage over people who start to exercise only in later years. And again, we come back to aerobic exercise. A study published in 2021 suggests that investing in rigorous exercise in your early years can benefit you in later years. In this study, people who biked or ran when they were younger did better as walkers in older age than people who just walked when they were younger. Specifically, those who just walked in their early years required more oxygen to walk at the same pace as those who ran or biked when they were young. The researchers speculate that a possible explanation for the difference in efficiency is better functioning mitochondria.[42]

Keep Going

Kyle's knee didn't get any better, but he's not someone who gives up easily. Searching for a solution, he reached out to Clint, another young man with FA who lived not too far from where Kyle got stuck. Clint's father managed a local machine shop in Odessa, Texas, where a solution emerged. Recalling his engineering classes, Kyle understood that there was a way to lighten the load on the injured leg. By situating the pedal closer to the center, his injured knee had a much shorter distance to move; that leg was just there to help the pedal come around.

Kyle got on the road again, and eight hundred miles later, almost against all odds, he reached downtown Memphis, where the National Ataxia Foundation conference was held. After the cheers, backslapping, and high fives from supporters at the finish line, he hopped on a bus to find a barber for a much-needed shave. A woman in a neighboring seat noticed the National Ataxia Foundation wristband on his hand. "Hey," she asked, "are you going to see Kyle speak today?" He shook her hand, smiled, and introduced himself. Slightly embarrassed, she laughed and introduced him to her twelve-year-old daughter, who had a different type of ataxia. "We've been reading your blog every day," she said, "and we really admire what you've done for our community."

"That was a moment where I realized that I was making a difference," Kyle told us.

It's hard to tell for sure what's happening in anyone's body. When we asked Kyle if he thinks that his ongoing physical activity has anything to do with the slow progression of his disease, he said that he believes it does, although he doesn't have any medical evidence to back that up. Years later, research is now starting to support his belief. In 2020, Zhen Yan from the University of Virginia School of Medicine found in a mouse model of FA that a four-month program of long-distance running at a young age prevented the symptoms of Friedreich's ataxia and improved the mitochondrial function of these mice.[43] This doesn't mean, of course, that exercise is a magic bullet in treating FA, and while it is possible that exercise can slow down its symptoms, there are still many unknowns, and there are FA patients who live beyond forty without exercising. We cannot emphasize this enough: Not all is in our control and an intervention that works for one patient may not help another patient as much (or at all).

David Lynch, a neurologist who specializes in FA at Children's Hospital of Philadelphia, pointed out to us that exercise creates new synaptic connections in our brains. "Exercise is the challenge to keep your synapses tuned up, so they continue to learn new things," he said. He also emphasized the value of exercise in reducing depression, which is experienced by many, including patients with chronic diseases. Indeed, there is evidence that exercise can lift our spirit, and we have already touched upon the link between depression and mitochondria in chapter 4. Exercise can also reduce psychological stress, which we'll discuss in chapter 10.

Eighteen years after Kyle completed his ride, people from all over the world regularly gather for such ride events to help fund treatments and maybe a cure. Each participant, sick or healthy, activates his or her mitochondria and produces more mitochondria and ATP. And cash! To date, these annual rides have raised over $10 million for the Friedreich's Ataxia Research Alliance (FARA), which was founded by Keith's parents, Ron and Raychel Bartek. As we saw in the previous chapter, this research into FA is starting to pay off.

●　　●　　●

We've been humbled by Kyle's incredible stamina and have been inspired to push ourselves a bit harder, especially since working on this chapter made us finally understand the "why" of fitness. We obviously had read about the value of exercise before, and have experienced its benefits over the years, but reading study after study that shows what happens to our mitochondria was more convincing to us. Looking back at the past four decades, we've always been engaged in physical activities—some years more than others—but the evidence we've seen here convinced us to ramp up both the length and the intensity of our exercise sessions. To take things up a notch. We hope it helped you, too. After talking to experts and reading the literature, here are a few guidelines.

Make it part of your life. When you engage in endurance exercise, you increase mitochondria content and improve their function. You also stimulate autophagy and mitophagy, which clear junk from your cells to rejuvenate them. You likely stimulate the release of myokines, which can affect the brain and other organs. And the same is true for the secretion of mitochondrial-derived peptides ("little treats") like humanin and MOTS-c, which have been shown to have various benefits. Incorporating rigorous activities (appropriate to your age and ability) into your daily routine can help. Daria walks at a fast pace to work and back every day and uses the stairs almost exclusively in her building. Emanuel exercises after his short daily nap. Such links to existing routines can be effective in turning physical activity into a habit: a stretching routine while watching TV or a walk after dinner are two other examples. And do something that you like! Your mitochondria don't really care how you get there as long as you get your pulse up: dance, hike, take a rigorous walk, swim, ride your bike, play basketball or another game. Finding an activity that you truly like increases the chance that you'll stick to it. "You've got to find what you enjoy because if you don't enjoy it, you'll never do it," Mark Tarnopolsky, who's been studying mitochondria and exercise for many years, told us. Tarnopolsky is a professor at McMaster University in Canada and has extensive experience in treating patients with mitochondrial disease. He is also an endurance athlete who competed internationally in sports such as winter triathlons and ski orienteering. When he was a competitive athlete, he focused exclusively on endurance training. He thinks about it differently now, which leads us to the next point.

Use a range of activities. It isn't only about endurance exercise. When Tarnopolsky was training, he focused on endurance but didn't think about weight training. He says this is typical of athletes, who tend to focus on only one type of exercise. "But for lifespan and health span, ideally you want a combination—a mixture of endurance and resistance training," he said, advice that is relevant to athletes and nonathletes alike. Resistance exercise refers to activities that use weights or body weight to strengthen your muscles. Building muscles is important because of sarcopenia (muscle loss), and building muscle early in life will give you reserve tissue that will help in later years. There are also important activities that are designed to improve your balance. Stretching activities are performed to minimize damage to your muscles. Doing something new may help you explore new movements. Whether you prefer a skateboard park, a climbing wall, your backyard, or a yoga mat in your office, the main point is that moving is good. Sitting for hours in front of screens (whether at your desk or in the living room) is not healthy, and experts recommend getting up and moving at least every thirty minutes. You're likely to get a nice thank-you note from your future self for engaging in all these activities.

Be mindful of the risk of overtraining. Pushing yourself a bit can induce mitohormesis, but be careful not to cross the line. We asked Nadège Zanou, the specialist from the University of Lausanne in Switzerland, how we can tell when we've crossed the line from beneficial exercise into damaging territory. "Listen to your body," Zanou said. People should feel well after exercise, not depressed or in pain. These are signs that something is wrong. If you notice, for example, that you become sick any time someone around you is sick, this may be a sign that your immune system is compromised as a result of overtraining. Some researchers talk about the 80/20 rule in this context. Do 80 percent of your exercise in Zone 2 and 20 percent in a higher zone. Recall that when exercising in Zone 2 (60 to 70 percent of your maximum heart rate), you can still talk, but you may need to stop to catch your breath. In higher exercise zones, your heart rate is even faster, while your ability to talk decreases further.

Consider making it social. By finding a workout buddy, you increase the chance that you'll stick with it. Peer pressure can be powerful, and you can use it to nudge yourself to action. Whether you work with your buddy online or go with her to the gym, it can help. In our case, perhaps the most important decision we made (the moment we could afford it) was to hire a personal trainer. We always viewed this as the fine we pay for our laziness, but we made it a top priority in the family budget. We used the help of personal trainers for over a decade and never regretted it. We realize that many people cannot afford this luxury, but a variety of free or inexpensive alternatives exist these days, from training apps to joining a group at your local community center to playing soccer with a team. The key word here is *accountability*. Having to explain to someone why you didn't show up makes you think twice before you decide to stay home.

Use it or lose it. While you can significantly help your mitochondria through exercise, the benefits from doing so can slip away if you stop exercising. Studies show that participants improve their mitochondrial capacity significantly after six weeks of aerobic exercise training. But if after those initial weeks participants stop exercising for four weeks or more, their mitochondrial capacity can decrease substantially (more so for elderly participants who return almost to pre-exercise values). The bottom line: You reap the full benefits of exercise only if you make it into a routine.[44]

CHAPTER 7

Feeding the Life Machines

In Greek mythology, Icarus and his father, Daedalus, were imprisoned on the island of Crete. The father, a talented craftsman, built them wings made of bird feathers and beeswax that would allow them to fly and escape. What was Daedalus's warning to Icarus before they took off? If you said that Daedalus warned Icarus not to fly too high, you are correct, but only partially. He also warned him not to fly too low because the feathers may get wet and heavy, and he would fall. Flying too close to the sun or flying too close to the seawater were both risky. Here is the father's warning as the Roman poet Ovid wrote it two thousand years ago: "Keep to the middle, Icarus. Follow my counsel: If you go too low, the sea mist will weigh down your wings. Too high, the heat will burn them up. Fly in between!"[1] We all know how the story ended: Icarus flew too close to the sun, the wax melted, and he fell to his death, but we should also remember what Daedalus understood—flying too low can be dangerous, too.

What diet is best for our mitochondria? This was one of the biggest questions we had while researching this book. Is it a low-carb diet? High fat? Low protein? Maybe something else? Opinions about diets are not hard to find, but we wanted to hear from a scientist who studies nutrition and has a deep knowledge of mitochondria. One such scientist is Alicia Kowaltowski, a professor of biochemistry at the University of São Paulo in Brazil, who has been studying energy metabolism and mitochondria for over thirty years. When we spoke to her, she pointed out right away that a single unifying dietary recommendation does not exist, because everyone has different dietary needs. She also believes that consuming too little or too many carbs, fats, or proteins is unhealthy. "Overall, the best scientific consensus is that diets that include all three groups in moderation are best," she argues.[2]

Some people find the idea of moderation disappointing; they seek extreme measures like eating only cabbage soup or only grapefruit. Yet research consistently shows that neither excess nor deficiency in biology is the way to optimal health. In this chapter, we'll cover several topics related to nutrition and mitochondria, and after poring over review articles and reading the opinions of additional experts, it seems that in many cases, taking the middle road is the best when it comes to feeding the life machines. As we mentioned before, scientists refer to this idea as the Goldilocks principle, named after the little girl who liked her porridge at just the right temperature—not too cold and not too hot.

Many people are confused about what to eat. According to one study, 80 percent of consumers say they came across conflicting information about food and nutrition and as a result, a significant portion of those doubt their choices.[3] But there *are* principles that many experts agree on: eating lots of vegetables and fruits that provide fiber and micronutrients (vitamins and minerals); drinking water and minimizing sugar intake; getting enough healthy proteins from fish, chicken, or other sources, and enough healthy carbs (whole grain); avoiding processed food as much as possible, and especially ultra-processed foods (e.g., packaged snack products, sugar-sweetened beverages, hot dogs, chicken nuggets).[4] When it comes to mitochondria and nutrition, one thing is clear: We saw in chapter 1 that mitochondria participate in assembling many of the building blocks of the cell. So it makes sense to give them the very best raw materials for this important job: the right food, at the right amounts, and with the right nutrients. Much of what you eat eventually will reach your mitochondria and will either help them or hurt them, and as we'll see next, giving them too much, or too little food, can lead to trouble.

What Eating Too Much (or Too Little)
Does to Mitochondria

Machines need fuel to operate properly, and the life machines are no exception. However, flooding them with fuel is not helpful. Recall that throughout the day, mitochondria switch between the two main fuel sources that come from the food you eat—fatty acids (which come from breaking down fat) and glucose (which comes from breaking down carbohydrates). Overeating confuses the mitochondria. If you repeatedly eat gigantic cheeseburgers with supersized portions of fries, your mitochondria may lose the ability to switch between the sources, a phenomenon known as metabolic inflexibility,[5] which occurs in people with obesity or type 2 diabetes. When we talk about sleep in chapter 9, we'll see another facet of this problem. Normally at night, mitochondria switch from metabolizing glucose to metabolizing fatty acids, but when you suffer from metabolic inflexibility, this important switch may falter. Consistently eating too much is associated with another problem called leptin resistance, in which people constantly feel hungry even though they have already eaten enough. Leptin is a hormone produced in fat tissue, and its job is to signal the brain that we had enough food to eat. The name leptin derives from the Greek word *leptos*, which means slim, so we can think of leptin as the hormone that keeps us slim (or at least slim-ish). But with obesity, this signaling mechanism can be broken. When people with obesity say that they are hungry, they are not making it up; it's the leptin resistance that makes them feel constantly hungry.[6]

Overeating might be the biggest health problem of our century, and it is associated with obesity and metabolic syndrome, a cluster of conditions that increases the risk of diabetes, stroke, and heart diseases.[7] A person with three or more of the following has metabolic syndrome: high blood pressure (over 130/85), high blood sugar (higher than 100 mg/dL), excess body fat around the waist (waist circumference of over thirty-five inches for women, over forty for men), high triglycerides (over 150 mg/dL) and low HDL, also known as "good cholesterol" (under 40 mg/dL). Importantly, there are thin people who suffer from metabolic syndrome and there are those with high BMI (body mass index) who don't. Still, a large percentage of those who are classified as having metabolic syndrome are overweight or obese, which brings us to caloric restriction. We'll discuss a big caveat in a moment, but there is evidence for the benefits of caloric restriction on our health, and studies also show that restricting calorie intake helps mitochondria. For example, caloric restriction has been shown (in mice and rats) to prevent age-related loss of mitochondrial oxidative capacity and efficiency, increase antioxidant scavenging, and minimize the damage to DNA and proteins from oxidative stress.[8] More research is needed before we know the exact impact of caloric restriction on mitochondria in humans, but many scientists agree that monitoring our calorie intake is a good thing. Most people in the Western world have access to much more food than they need, so for many people (including us) it means that we need to watch how much we eat.

But the goal of caloric restriction shouldn't be to make us look like stick figures.

Eating too little is bad, too. Mitochondria need glucose and fatty acids to produce ATP, and these come from the food you eat. Since your mitochondria are working 24/7 to supply ATP to your heart, brain, liver, and other organs, you need a basic number of calories to keep you alive, and on top of this you need to fuel your daily activities. As a very general guideline, adult females are likely to require between 1,600 and 2,400 calories a day, while adult males need 2,000 to 3,000 calories per day. How many calories you need also depends on your age, your activity level, and how much you exercise.[9] If you're constantly at a calorie deficit, you may feel tired, easily irritated, and cold. You may also suffer from muscle loss (sarcopenia) if your diet doesn't include enough protein (plus the physical activity that is needed for muscle building).

As you know by now, mitochondria are involved in much more than energy transfer, and certain micronutrients are essential for the life machines' many functions. Lack of such nutrients can accelerate aging. Starving yourself continuously and hoping to get all the necessary nutrients through supplements is not the way to a healthy body.[10] There's plenty of evidence about the heavy impact of malnutrition on brain development in childhood,[11] and long-term malnutrition in adults can cause problems, too, as is the case with certain eating disorders. For example, in a study conducted in Spain, researchers examined the white blood cells of forty young women, twenty of whom were patients with anorexia and twenty who served as controls. The mitochondrial functions of the cells of the anorexic patients were impaired, and those mitochondria produced more ROS.[12] Since some nutritional recommendations are based on studies in rodents, it's worth noting that calorie restriction experiments in mice or rats are not likely to show these problems, because lab animal's food usually provides all the nutrients they need, including minerals and vitamins.[13] This isn't always the case when humans drastically cut down on calories and in doing so miss important micronutrients.

A key to healthy mitochondria can therefore be found in the advice of an ancient father: "Keep to the middle, Icarus." Too much food can confuse your mitochondria, while too little can starve them from important nutrients.

Your Mitochondria's Favorite Micronutrients

Think about what you had for dinner last night. As you're trying to remember, we'll tell you what we had: a piece of salmon, broccoli, potatoes, a salad, and some ice cream. Now, let's analyze our dinner from a very rudimentary nutritional perspective, and you can go along, doing the same regarding your dinner. The salmon contained proteins and some fats; the broccoli contained some proteins, carbs, and lots of fiber; the potatoes contained mostly carbs, a little protein, and a tiny amount of fat. The salad (leafy green, cucumber, orange, and avocado) contained mostly fiber, some carbs, some protein, and fat (from the avocado); the ice cream contained sugars, fat, a little protein, and carbs.

How is this food converted into ATP that fuels our activities? The main sources for that are fat and carbs so we'll focus on them. First, they are broken down in your digestive system, carbs to simple sugars (glucose), and fat into fatty acids. Next, they travel in your blood to individual cells. In the final step before they enter the mitochondria to produce ATP, these things are converted into acetyl-CoA. This is that molecule that we likened in chapter 1 to a suitcase loaded onto the baggage carousel—the Krebs cycle—where it goes through a set of chemical reactions to produce NADH and $FADH_2$. Now starts the ATP production process, which we have already discussed. After the Krebs cycle come the electron transport chain, the proton gradient, and finally that amazing rotor (known as Complex V) that produces ATP.

Yet all these conversions don't happen by themselves; they need the help of some special micronutrients. This is critically important to understand, and to learn about the discovery of one of these micronutrients, let's travel back in time to 1886 and to the Dutch East Indies (today's Indonesia). There, a Dutch physician named Christiaan Eijkman was investigating a disease called beriberi that caused paralysis of the legs and other neurological problems among local residents.[14] The disease affected not only humans. Eijkman observed how some sick chickens there would frequently collapse and fall over when walking, and some became almost immobilized. Then suddenly the disease cleared up; the sick chickens recovered, and there were no new cases. Eijkman noted that the chickens got sick after they were fed with rice from the local military kitchen, and that they recovered once they were fed with rice from a different source. What was the difference between the two types of rice? Follow-up research revealed that the rice from the military kitchen was treated in a way that caused it to lack an ingredient that today we call vitamin B_1.

Vitamin B_1 helps in the conversion of pyruvate (a metabolite of glucose) into acetyl-CoA, and then it is involved in the production of ATP inside the mitochondria. Recall that the brain is a big consumer of ATP, so the lack of vitamin B_1 explains the neurological problems of the beriberi patients and the sick chickens. Today, beriberi is rare because most people get enough B_1 in their diets to prevent this disease. Still, it is important to make sure you get enough B_1, because a mild deficiency is associated with other problems, including fatigue and irritability.[15]

B_1 is just one of eight B vitamins that are crucial to proper mitochondrial functions. There is no life without mitochondria, and mitochondria cannot operate without B vitamins. We must make one thing clear right away. When we talk about vitamins, we refer to compounds that naturally can be found in certain foods, not to be confused with supplement pills that may contain vitamins, which we'll discuss later. A healthy varied diet should provide most people with the vitamins they need. Luckily, the quantities of most of the vitamins we need are tiny, and we get them from the food we eat. However, vitamin B deficiencies still exist, and to avoid them, we should make sure we eat a balanced and diverse diet. Below is the list of those vitamins and some of the foods that provide them.[16] Of course, this doesn't mean that you must eat every item on the list, or that you should start eating red meat if it is listed as a source for a particular vitamin (there are well-documented reasons to limit red meat consumption).[17] What this list means is that you should regularly eat food that provides you with each type of vitamin B. The word *regularly* is important here because vitamins in the B category are water-soluble, which means that they are washed out of our body regularly, as opposed to fat-soluble vitamins (A, D, E, and K) that are stored in our body for a longer period. (Incidentally, if you wonder why we said that there are eight B vitamins, and you see that the list goes to B_{12}, the answer is that some compounds that were initially believed to be vitamins were eliminated from the list.)

- **Thiamine (B_1)** is found in brown rice, lentils, peas, oranges, bananas, nuts, whole grain bread, and other foods.
- **Riboflavin (B_2)** is found in milk, eggs, fortified cereals, mushrooms, and plain yogurt.
- **Niacin (B_3)** comes in two forms—nicotinic acid and nicotinamide—which can be found in meat, fish, wheat flour, and eggs.
- **Pantothenic acid (B_5)** is found in chicken, beef, eggs, mushrooms, and avocado.
- **Pyridoxine (B_6)** is found in pork, chicken, turkey, some fish, peanuts, soybeans, wheatgerm, oats, bananas, milk, and some breakfast cereals.

- **Biotin (B$_7$)** is made by bacteria in the microbiome. It is also found in egg yolk, liver, legumes, sweet potatoes, nuts, and seeds.
- **Folate (B$_9$)** is found in broccoli; brussels sprouts; leafy green vegetables, such as cabbage, kale, spring greens, and spinach; and peas, chickpeas, and kidney beans.
- **Cobalamin (B$_{12}$)** is found mainly in meat, fish, milk, cheese, and eggs.

Again, although the quantities we need are tiny, deficiencies in B vitamins still occur, and vitamin B$_{12}$ (cobalamin) is worth some special attention. As mentioned, it is found mainly in meat, fish, milk, cheese, and eggs, so people who follow a vegan diet, for example, may suffer from a variety of problems, including anemia due to B$_{12}$ deficiency. Chris Palmer, a Harvard psychiatrist who specializes in metabolism, suggests that certain vitamins (not only B$_{12}$) should be routinely checked in patients with psychiatric and neurological disorders, because if they are low, then there is a clear treatment for these disorders.[18]

In addition to foods that contain B vitamins, mitochondria also love foods that contain antioxidants. Recall that most free radicals are made in the mitochondria, and the mitochondria also produce antioxidants that neutralize those free radicals (up to a point). Antioxidants can be found in blueberries, dark chocolate, artichokes, nuts, and many vegetables. Vitamin C (which, like B vitamins, is also water-soluble) is a good antioxidant, and it can be found in citrus fruit as well as in many other natural sources.[19] Vitamin E and other compounds are antioxidants as well and can be found in fruits and vegetables.

Mitochondria also need certain minerals like iron, magnesium, calcium, and zinc.[20] Iron is another example for the Goldilocks principle. If you have too little iron, you may feel unusually tired, short of breath, and weak. On the other hand, too much iron can be toxic and can damage important organs and your DNA. This is an example of why you should consult with your health-care professional if you are considering taking nutritional supplements. If a blood test indicates that you are low on iron, your doctor may prescribe supplementation. In some cases, iron deficiency stems from poor nutrition, and in other cases it may be the result of mitochondrial deficiency as we age. (It is one of the most common deficiencies in older people and may affect cognitive capacity and memory.)

The Goldilocks principle applies to magnesium, too. Magnesium supplementation improves cardiac functions and reduces insulin resistance in diabetic patients. Researchers from Brown University and the University of Minnesota showed that these benefits of magnesium supplementation in diabetic mice with cardiac dysfunction are due to improved mitochondrial structure and functions and reduced oxidative stress.[21] However, this shouldn't make us all take in large amounts of extra magnesium, because doing so can lead to irregular heartbeat, confusion, and even death.

To discuss the next nutritional ingredient that our mitochondria need, we'd like to pause for a second and focus on one of the most important things that makes our body work—membranes. These are thin and flexible layers that surround (and define) the cell and different organelles. Keeping things away from each other in the cell is important. For example, we saw that when calcium enters the cell, it is quickly taken up by the mitochondria so that it doesn't interact with other cell components, which may lead to the death of the cell. And as we saw, a mitochondrion has two membranes, a fact that enables the magic of ATP production. High-quality membranes, therefore, are crucial to the smooth operation of the cell. Membranes are made of lipids (fats), and you want to get the best fats to build your membranes. If membranes are made from low-quality fats, they don't keep things away from each other the way they should.[22] High-quality fats are those that come from olive oil, avocado oil, or certain fats found in fish for example.[23] Two fatty acids that differ in their chemical structure, omega-3 and omega-6, are both needed, but the ratio between them is important. In the typical Western diet, the ratio between omega-6 and omega-3 fatty acids is estimated to be around 20:1 in favor of omega-6, and it should be much more balanced.[24] Foods rich in omega-3 include sardines, salmon, chia seed, and walnuts. Foods rich in omega-6 include canola oil, almonds, sunflower seeds, eggs, and tofu. Omega-6 fatty acids are also found in processed food, which explains the unbalanced ratio.[25]

Two more points that we'll develop later in the book are important to mention here. First, there's one more thing that mitochondria love for you to eat—fiber. Mitochondria don't consume fiber directly, but eating fiber allows your microbiome to create chemicals that are very important to proper mitochondrial functions. You'll find more on this in chapter 8. Second, we have focused so far on nutrients that are good for your mitochondria, and we'll dedicate chapter 11 to things you should avoid as much as possible. In that chapter, we cover things such as pesticides and exhaust fumes, but also a few things that relate to nutrition, like fried and barbecued food, trans fats, and alcohol.

One thing that should be minimized is sugar. The problem with simple sugars like glucose and fructose is that they are absorbed very quickly. When eaten alone and in large quantities, they create a spike in glucose in our blood leading to the release of insulin, which over time can cause diabetes, cardiac diseases, dementia, and other chronic diseases. It's best then to minimize their consumption and avoid soft drinks or fruit juice. Some experts go all the way to recommend avoiding fruit (that contains fructose), but that advice ignores the fact that fruits provide important nutrients and fiber. In general, keeping sugar consumption to a minimum is certainly advisable, but should sugar be completely eliminated from our diet? When we directed this question to Alicia Kowaltowski, she said that, nutritionally, there's no reason to eat simple sugars, but she also pointed out that avoiding a birthday cake when everybody is celebrating someone's birthday can be socially isolating, and celebrating with everyone is part of living well, too. We like that.

When you think about reducing sugar consumption, perhaps you think about skipping ice cream or avoiding chocolate cake at the next party, but different types of sugars can be found in many savory foods because they help preserve food for longer (and sugar attracts consumers). So even when buying baked beans or ketchup, it's a good idea to look on the label for those hidden sugars of such processed products. Look for things that end with "ose" (e.g., fructose, sucrose, maltose, dextrose), things that say syrup (e.g., corn syrup, rice syrup), and obviously for ingredients that contain the word sugar (e.g., cane sugar, confectionary sugar). Try to avoid these foods or eat them less often.

Going back to the dinner you had last night, or to the meals you'll have tomorrow: It's a good idea to make sure that your meal contains ingredients that give your mitochondria the vitamins they need, some antioxidants, minerals, good fats, and fiber. Taking our dinner from last night as an example: We got lots of fiber, minerals, and vitamins from our salad (especially since we added that avocado). The salmon contained lots of B_{12}, vitamin D, iron, potassium, and omega-3 fatty acids. The broccoli boosted the meal with additional fiber and vitamins. The ice cream had mainly fat and sugar, so it didn't help our mitochondria much. (But it was so good, and we allow ourselves to indulge from time to time.) Except for the ice cream, none of this food was processed in a factory. A simple natural meal, just as our mitochondria like it.

Dietary Supplements and Mitochondria: More Isn't Necessarily Better

A healthy balanced diet should provide most people with the vitamins and minerals their mitochondria need. It is best to support our mitochondria through the food we eat because our body seems to have a way to deal with excess. Food provides nutrients in a form that our gut can absorb properly, which cannot always be said about dietary supplements, and the vitamins that you get directly from food come with many other nutrients. Having said that, there are specific cases where people suffer deficiencies. One such case was noticed by a British physician named Lucy Wills in 1928. Wills was invited to India to investigate a certain anemia among pregnant women that was associated with severe birth defects. She noticed that the cases were especially prevalent among the poor, and like Christiaan Eijkman, she suspected that this had to do with a nutritional deficiency. Her hypothesis was confirmed when she discovered that adding marmite (a cheap breakfast spread commonly consumed in the U.K.) to their diet cured poor women from this anemia within days. The compound in marmite that did the trick was initially named "the Wills factor," and today we know it as folic acid or vitamin B_9, which is widely prescribed to women during pregnancy, even to women who already enjoy a rich diet.[26] In many countries (including the United States), flour is fortified with B_9 so that women get it even before they find out they are pregnant because this vitamin is so important during pregnancy. Indeed, studies show that fortifying food with folic acid has been associated with a significant decline in the prevalence of neural tube birth defects.[27]

But the fact that B_9 supplementation benefits pregnant women doesn't mean that everyone should take these supplements regularly. And yet, researchers argue that many in North America are exposed to high levels of dietary folate from the combination of dietary supplements they take and from fortified food.[28] How many people are exposed is a subject of debate, and scientists don't know exactly how these high levels of vitamin B_9 affect people's health, but studies show increased association between excess vitamin B_9 supplementation and cancers, immune and cardiovascular diseases, and other health problems.[29] Our point is not that one should always avoid taking folic acid supplement. Our point is that we should avoid the fallacy of "more is better" and immediately swallow everything that is supposed to have beneficial effects. Supplementation is a decision that one should consider carefully and make together with one's physician because it's complex.

There are many problems with overdoing it with vitamin supplementation. For example, taking 200 mg or more a day of vitamin B_6 can lead to peripheral neuropathy (loss of feeling in the arms and legs).[30] There are also interactions between dietary supplements and medications. For example, vitamin K, which has been shown to help with mitochondrial functions,[31] might interact with blood thinning medications like warfarin (the generic version of Coumadin, which prevents excessive blood clotting).[32] Again, this is not something that you want to manage on your own. You should always seek medical advice when considering new dietary supplements; and when prescribed a new medication, always list *all* the supplements you take, since there may be reasons you should avoid them due to your specific medical condition.

There are instances when vitamin supplementation *can* be very beneficial in addressing specific deficiencies. Here's an example: Anu Suomalainen is a Finnish physician scientist who's been researching mitochondria for many years. In a 2020 pilot study, she and her team recruited five patients with mitochondrial myopathy (a disease that causes muscular problems) who they found to suffer from NAD+ deficiency. NAD+ is an essential regulator of multiple cellular functions, and based on experiments in mice, the researchers learned that this mitochondrial disease led to secondary NAD+ deficiency. This could be addressed through supplementation with NAD+ precursor niacin, a B_3 vitamin. After ten months of treatment with niacin, these patients experienced significant improvements and some even partial reversal of their condition. "It was remarkable to see," she told us. Some patients who previously could walk only a few hundred yards could now walk several miles without the cramps they used to have. But Suomalainen stressed that the positive results applied to certain humans who had disease-caused NAD+ deficiency[33] and that not everyone should be treated with niacin.

The above examples demonstrate that supplements can help address deficiencies in specific patients or specific populations (such as pregnant women). Other cases are less clear-cut. A 2023 large clinical trial from Columbia University and Harvard Medical School showed that taking a daily multivitamin supplement can slow age-related memory decline in older adults.[34] However, a 2013 large study from Harvard among older men did not show such a difference among those who took multivitamins.[35] Many experts agree that supplementation is needed only when a clear deficiency is detected, yet Americans take dietary supplements whether they need them or not; more than half of adults can be qualified as regular users.[36]

What we write here is a snapshot of the current knowledge, which is why consulting with your physician is a must. New studies keep coming, and your health professional hopefully is on top of things. Reliable sources such as the Mayo Clinic or Harvard School of Public Health[37] update their recommendations based on the latest research, so it's a good idea to use them in addition to your health professional. Consider the case of antioxidant supplements. In the 1990s, research established the link between free radicals and several pathologies such as atherosclerosis and cancer. Additional research showed that people who got fewer antioxidants from things like fruits and vegetables were more susceptible to getting these diseases, which led to sweeping recommendation of antioxidants in the media and to exaggerated claims from dietary supplement companies and food companies. Yet additional research and clinical trials showed that the initial excitement was not justified. Eliminating every free radical from every part of the cell is impossible, and it can also interfere with important signaling. As we pointed out earlier in the book, mitochondrial ROS play a role in signaling, so totally eliminating them causes problems like decreased benefit from exercise for example.[38] Goldilocks strikes again: consuming too little antioxidants is bad, as is consuming too much. We need just the right balance, which is usually provided through following the right diet.[39]

As mentioned, NAD+ is essential to the operation of the mitochondria (when combined with two electrons and a proton, it turns into NADH, which you may remember is critical for ATP production). Given its importance, some people consider taking what are known as "NAD boosters," which promise to enhance mitochondrial functions. Nicotinamide riboside (NR) and nicotinamide mononucleotide (NMN) are derivatives of niacin (vitamin B_3), which we met already. Those supplements have shown some promising results in mice, but at least so far, significant positive results in healthy humans have not been demonstrated.[40] This can change, of course, which is why you should stay tuned.

Another mitochondria-related supplement that gets a lot of attention is coenzyme Q10 (CoQ10). CoQ10 is an electron carrier that is found in your cells, especially in the electron transport chain of mitochondria. Your body produces it naturally, and it is also found in some foods (meat, fish, nuts), but in certain cases, research suggests that a supplement may be beneficial. For example, Beatrice Golomb, a professor at UC San Diego, demonstrated that CoQ10 can help veterans who suffer from Gulf War illness, a chronic multisymptomatic problem that affects many Gulf War veterans. (Symptoms include fatigue, headaches, muscle aches, sleep problems, and cognitive impairment.) In 2014, Golomb published the results of a double-blind randomized controlled trial demonstrating that coenzyme Q10 alleviated the symptoms of Gulf War illness significantly and improved physical function.[41]

Oral supplementation of CoQ10 shows benefits in other indications, including female infertility (demonstrated in five randomized-controlled trials)[42] and moderate to severe heart failure.[43] CoQ10 may also help reduce cholesterol in diabetics, decrease the frequency of migraine headaches, and ease muscle pain experienced by some people who take statins (statins decrease the synthesis of CoQ10).[44] However, it doesn't seem to improve symptoms of Parkinson's disease patients, for example,[45] and there are mixed results regarding CoQ10's effect on reducing blood pressure.[46] Again, this is a snapshot of the research on the potential benefits of CoQ10 at one point in time, but things can change.

A final point about dietary supplements: safety. Many people who take supplements believe that manufacturers must prove to the FDA that their products are safe before they can be marketed, or that the FDA tests supplements. This is not the case. By law, the FDA cannot review or test dietary supplements before they are sold. They can only start to enforce quality standards after the supplement is available on the market, and with tens of thousands of different products, this is not a trivial task. C. Michael White from the University of Connecticut argues that consumers are taking real risks when they use supplements that were not independently verified by reputable outside labs. For example, two nonprofit groups that provide certification in the United States are NSF and USP.[47] Unfortunately, White also reports that it's not uncommon to find bacterial or fungal contamination in supplements. There have also been multiple cases of the presence of arsenic, lead, and mercury that exceeded the minimum allowed. All this means that it is our responsibility as consumers to verify the product quality and safety: In addition to testing by independent labs, this can be done by assessing the manufacturer's reputation and consulting with one's health-care provider and pharmacist. If you consider taking a dietary supplement, you clearly do it to improve your health or performance, so the extra time invested in research is definitely worth it. If you're still not convinced, here are four additional reasons to do your homework, consult with professionals, get supplements only from a source you fully trust, and take claims with a grain of salt:

1. What you read on the label is not always what's in the pill you take. It can contain, for example, a prescription drug you may be allergic to or try to avoid because you're concerned about its interaction with another drug that you use. The FDA found this to be the case in some supplements that claim to help with sexual performance, weight loss, and muscle building.[48]

2. Dietary supplements don't always contain the quantity specified on the label. For example, a 2017 study found that the content of melatonin in melatonin supplements varied widely from what the label said. In more than 70 percent of the products tested, melatonin content wasn't within a 10 percent margin of the claim on the label. The actual content was as high as 478 percent more than the concentration on the label.[49]

3. Makers of dietary supplements can say pretty much anything, as long as they don't make claims on benefits for specific diseases. So you frequently see vague claims such as "improves brain performance" that are not necessarily backed up by solid research. Another word the industry uses frequently is *natural*, which again doesn't mean that it is safe. "Snake venom is natural. That doesn't mean it's good for you," says Alicia Kowaltowski.

4. Relying on questionable online information sources and the availability of supplements for sale on the web raises serious concerns, too. For example, a substance that has been banned in the United States and in other countries is sold online as a weight-loss drug. These pills contain a molecule called DNP, which tricks the mitochondria into dissipating energy as heat, which means that more calories are burned as heat instead of producing ATP. The heat produced by these pills can be mild, but a slight excess of DNP can literally cook you alive. There have been multiple reports of death as a

result, and people often find out about such pills in online forums.[50] As we point out later, using medical sources and your physician to verify any claims you come across in the media or social media is a good idea. See "Think Like a Scientist" at the end of this chapter.

Fasting and Your Mitochondria

It seems that everyone is talking about fasting lately, and even if you haven't been part of this trend, in a way, you've been practicing it since the day you were born. We all fast when we are asleep, which is a good thing, because this is when our mitochondria focus on recovery. As explained earlier, during the day, they work hard to match your energy demands, and during the night they switch to recovery mode, which is why raiding your fridge at midnight is a bad idea. We'll discuss this further in chapter 9, which is dedicated to sleep. Fasting has become a hot topic in recent years because some scientists believe that extending the fast period beyond the time that we sleep can be helpful. The main idea is that you narrow down the window of eating to eight to ten hours, which means that you fast for fourteen to sixteen hours. The term used for this practice is *time-restricted eating* (and in animals, *time-restricted feeding*). A similar popular practice is intermittent fasting, where on some days of the week you eat as usual and on others you fast or restrict your calorie intake significantly.[51]

Can mitochondria benefit from these fasting practices? The jury is still out on this question, and these practices are not without risk. The main potential benefit may come from inducing the cellular recycling programs—which we discussed in the first part of the book—that eliminate damaged parts of the cell (autophagy) or damaged parts of mitochondria (mitophagy). However, these associations have been established mostly in animals, and in periods of fasting of several days, which require medical supervision and certain precautions. Even short fasts can be risky for certain people, for example, for those who suffer from type 1 diabetes, are pregnant, or are prone to eating disorders. And it's not clear at all that mitophagy or autophagy happen under shorter fasts.

The research on a variety of fasting practices does show some promise, though. A 2017 Harvard study of the mitochondria in *C. elegans* (one-millimeter worms), showed that the worms remained more "youthful" after a fast, and that the fast even expanded their lifespan.[52] Studies in other animals showed that time-restricted feeding can prevent or attenuate the severity of some metabolic diseases, such as glucose intolerance (inability to bring back blood glucose to a normal level—a sign of prediabetes) and dyslipidemia (abnormal level of lipids in the blood). But while there are reasons to be optimistic about this line of research, there are not enough longitudinal studies of such fasting practices in humans, so we don't really know their long-term effects. And there is another factor to consider—genetics. Many studies are done on mice that are genetically identical to one another, so they ignore the fact that we, humans, are not all the same. There is a 2024 study that is an exception to this; it used 960 female mice with very different genetics.[53] The mice were divided into five groups and were placed on the following diets throughout their lives. One group was an "all you can eat" group. Two groups were fasting groups: in one, the mice fasted for twenty-four hours each week, and in the other forty-eight consecutive hours each week. The last two groups were the caloric-restriction groups: in one group, the mice ate 20 percent less than the amount eaten by the "all you can eat" group, and in the other group, the mice ate 40 percent less than that eaten by the "all you can eat" group.

There were two important insights from this study: First, that genetics matters more than diet for the lifespan of mice. To be clear: This does not mean that diet didn't matter, but that genetic differences between the mice played an even more important role, suggesting that benefit of dietary restriction (both intermittent fasting and calorie restriction) in humans may be highly individualized. The second insight is that while the mice that were put on the more severe caloric restriction lived longer, they were not necessarily healthier; there are reasons to believe that they were more susceptible to infections, for example. (To that end, the researchers assessed the robustness of the rodents' immune response, and such blood test analysis can be done in humans.) Another surprising finding was that among the mice on a calorie-restricted diet, those that had the lowest weight were more likely to die earlier relative to other mice in the same group of calorie-restricted diet. There are good reasons to believe that eating less is good for longevity and health, but this massive study reinforced our belief that more research is needed and that taking extreme measures can be problematic.

Even short-term studies in humans are not without problems, because they usually don't take into consideration what type of food people ate when they were not fasting. In addition to the fact that we don't know enough, focusing too much on fasting—on *when* we eat—might shift the attention away from the most important issue—*what* we eat. For example, some people who practice time-restricted eating achieve the fasting window of fourteen to sixteen hours by skipping meals, mainly breakfast, which can lead to them not eating enough proteins. Our body doesn't store amino acids. We get them from the food we eat, so by skipping breakfast, we can starve the body from important amino acids (the building blocks of proteins and some important neurotransmitters).[54]

Personally, after weighing the risks and possible benefits, we decided to adopt a fourteen-hour version of time-restricted eating and have managed to stick to it for several years. We never fast for more than fourteen hours, and sometimes just twelve or ten. So far, the only tangible measurable benefit for us is that it put an end to Emanuel's endless snacking while watching the evening news.

There's More than One Healthy Diet

Many discussions about the ideal diet end with unequivocal recommendations. The one and only diet is X, or the best diet is Y. But these recommendations ignore the fact that as humans evolved in different parts of the world, they consumed certain foods for thousands of years and were genetically selected to benefit from these foods and sometimes lost the ability to benefit from other foods. For example, we often hear about the Mediterranean diet as the ideal diet to adopt. It happens to be our favorite diet, too, but does it mean that it is perfect for everyone? Is it wise to recommend it to someone whose ancestors had a fish and seaweed-rich diet for millennia? Consider this simple fact: Almost 95 percent of Europeans are lactose tolerant, meaning that they can consume dairy without any issues. In contrast, more than 70 percent of East Asians are lactose intolerant. There is a genetic explanation to this dramatic difference: Lactose intolerance is a common trait in all mammals after weaning from nursing, but around nine thousand years ago Europeans began inheriting a genetic mutation that enabled them to digest lactose throughout their lives.[55] This mutation didn't become a common genetic trait in East Asia (where it provided no advantage, because milk was less commonly consumed), explaining the high prevalence of lactose intolerance among people with East Asian roots. This and other examples demonstrate that genetic differences affect responses to specific diets. There's an old Dutch and German proverb: What the farmer doesn't know, he doesn't eat. To be clear, we love trying new foods, but maybe there's some truth in this proverb, and being adventurous with food is not always advantageous.

To better understand what we're talking about, let's travel back in time to Sardinia, an island near Italy. For centuries, the island was surrounded by marshes, which means that there were lots of mosquitoes and therefore malaria. Thousands of years ago, some people there had spontaneously developed a genetic mutation that impairs the functioning of a certain enzyme called G6PD. This enzyme helps red blood cells (that don't have mitochondria) to produce NADPH, an important antioxidant. Having this deficiency wasn't great for overall health, but it was advantageous against malaria because it increased ROS in red blood cells and killed them along with the parasite that causes malaria that lived inside these red blood cells. So a malaria-infected person who had G6PD deficiency was more likely to reach childbearing age and have kids. Since this is an inherited mutation, that person's kids were likely to have the same mutation and to survive malaria that killed their neighbors. Over time, more people with reduced G6PD activity survived, and those without the mutation died of malaria. What does all this have to do with food? When fava beans, which originated in Iran and other Eastern Mediterranean regions, were brought to Sardinia, the advantage of those with G6PD deficiency became a disadvantage. People with G6PD deficiency who eat fava beans experience breakdown of their blood cells (a phenomenon called hemolytic anemia) and may even die. Today, 10 to 15 percent of the population in Sardinia have this deficiency. Like lactose intolerance, fava bean intolerance is not an allergy, but an example for a mismatch between genetics and a certain food. The advantage of G6PD deficiency that some people inherit does not match a diet that includes fava beans, lentils, peanuts, and peas. (Note that this problem is not restricted to Sardinia, and it is estimated to affect about 400 million people worldwide. Being informed about this is important because some common drugs may induce hemolytic anemia in people with this deficiency.)

When we asked Anu Suomalainen, the Finnish researcher, about the Mediterranean diet, she laughed and said: "It may be that the Mediterranean diet is good for Mediterranean people," but she questioned whether it is necessarily the perfect diet for everyone. She has a point: What's true for people in Greece or Italy is not necessarily true for people in Finland, who for generations have fed on a completely different diet. Much more research should go into testing the effects of certain diets on different populations. And even what's good for another person from your country isn't necessarily good for you, so it makes sense to adopt a skeptical point of view and to assess carefully any new diet, as we discuss next.

Think Like a Scientist

The debates about things like diets and exercise regimens won't end any time soon, and in the coming weeks, months, or years, you're likely to come across new claims regarding mitochondria and how to help them. Science keeps changing, and as you encounter new claims, we encourage you to adopt the healthy skepticism of a scientist by following a few simple guidelines. First, pay close attention to the credibility of your source. If you come across some claim in the media or social media, check it against medical sources. Your doctor's online portal may include links to such credible sources of information. MedlinePlus (from the National Library of Medicine) is a good site that offers easy-to-read health information. The Medical Library Association maintains a list of top health websites.[56] What are these sources saying about the new claim? You may come across more than one opinion, which isn't necessarily bad.

You may want to look at the journal article that is behind the claim that you are evaluating (or at least read the abstract, the summary). If you don't have any scientific background, reading a scientific article can be tricky, but it has gotten somewhat easier in recent years because you can quickly look up terms online, and some journals now include a visual abstract that makes things more accessible. Evaluate the actual claim in the journal article. Is it exactly what was claimed in the media or social media source? (There are often discrepancies.)

When evaluating the study, here are some questions that you should consider:

- Is the article an opinion piece, or is it based on experiments? There is nothing wrong with opinions, but remember that they are not facts.
- Is the article an observational study? In this type of study, scientists report a correlation between two phenomena. Suppose, for example, that you came across a study that claims that people who eat bananas sleep better. The authors have not proven that eating bananas is the cause of good sleep; it is just a correlation.
- Was the study done on a large scale or was it a small pilot study?
- Was it a randomized, double-blind, placebo-controlled trial, which is considered the gold standard in interventional studies? In this type of study, participants are randomly assigned to an experimental group (that is treated) and to a control group (that is not). Neither the participants nor the researchers know who is getting the treatment and who is not.
- Are there other independent studies that support the same claim?

Now, try to assess how relevant the claim is to your situation: Was the study conducted on humans or on animals? We can learn a lot from animal studies, but a lot more needs to be done before we can apply that knowledge to humans (and this cautionary note applies to the animal studies we have cited in this book). If the study was conducted on humans, were the participants in this study similar to you in their age or medical condition? If the claim seems like it has merit and is relevant to you, you still may want to wait for further evidence or get your health professional's opinion. If you decide to adopt a new practice, proceed with caution. Start slowly and evaluate any side effects. Sometimes you want to implement several changes simultaneously, and often there are good reasons for that. However, if it's important to you to find out exactly what intervention works, you may want to make one change at a time. (Although the interaction between different interventions may have a positive effect, too, that you won't be assessing by isolating the interventions.) And ideally, evaluation should be objective: a change in clinical lab results, a change in unbiased measures like blood pressure or cholesterol level. If you change your diet dramatically, it's a good idea to follow up with thorough blood tests to assess your micronutrient needs and adjust according to your health-care provider's recommendations. Try, evaluate, reassess, and stay tuned, because there's always a new study that may shed new light on things or how they relate to one another.

Thinking like a scientist will become somewhat easier with the emergence of better measurement tools. Measurement and science go hand in hand. There are currently many tools to monitor behaviors that may affect our mitochondria; for example, apps to assess our physical activity, nutrition, sleep, exposure to air pollution, and stress level. There are also devices for continuous glucose monitoring or assessing metabolic flexibility. But unfortunately, there is nothing similar when it comes to directly measuring mitochondrial functions. We can only dream of a wearable device that would easily monitor mitochondrial functions in real time. Until we have such a device, we must rely on proxies such as VO_2 max, blood tests for lactate, oxidant levels, inflammatory markers, or other indirect biomarkers. Other current methods (such as skin biopsies) are expensive and impractical for continual monitoring, but it is always possible that new tools will emerge. A breakthrough in this area can be revolutionary because it will provide direct feedback on therapeutics and lifestyle changes. Imagine being able to say precisely how a certain medication affected mitochondria in your heart, or how changing your diet affected mitochondria in your brain. Currently, the lack of accessible measurement tools may be the biggest obstacle to knowing how changes in our nutritional habits can improve mitochondrial functions. So for now (and likely for a long while), uncertainty is part of the deal, and we hope you develop a commonsense approach to evaluating claims that will allow you to fly not too high and not too low. Have a safe and enjoyable flight.

CHAPTER 8

You, Mitochondria, and Your Gut

Let's talk about two ancient tribes: One consists of the 10 million billion mitochondria that originated from bacteria; the other includes the bacteria that live in your gut microbiome, a complex ecosystem of trillions of microbes living in colonies in your intestines. Mitochondria and gut bacteria have been living side by side inside humans forever, and yet, only in recent years has research begun to address questions such as: How do these two tribes help each other? Do they sometimes compete or fight? Do they communicate somehow? And most important, What can we do to help these two tribes live in harmony?

Both these tribes were passed on to you from your mother; you inherited your mitochondrial DNA from her, and you got your initial microbiome from your mother as you passed through the birth canal. (Unlike your mtDNA, your microbiome has developed further based on the food you eat and other factors. It is still evolving and changing.) A quick example of the cooperation between the gut bacteria and your mitochondria is the following: We saw in the previous chapter how important B vitamins are for the proper operation of our mitochondria, and it turns out that bacteria in your gut provide certain B vitamins.[1] On the other hand, other bacteria sometimes seem to compete with our mitochondria. For example, researchers from Yale University showed that certain bacteria in our microbiome may grab vitamin B_{12} as it passes through the digestive tract before the mitochondria in our tissue have a chance to get it.[2] As we'll discuss, there are things you can do to help these two tribes cooperate.

An important touchpoint between the two tribes is the health of the gut barrier. Our gut barrier is an essential part of our digestive system, and its role is to keep the microbes that reside in our gut away from our bloodstream. Mitochondria are instrumental in this task by supporting the gut lining cells that make that barrier. Maintaining that separation is key because bacteria and bacterial toxins can be dangerous if they enter the body. In the same way that there are helpful and harmful organisms in a garden, the microbiome, too, has both "good" and "bad" bacteria. (We put them in quotation marks because the lines between good and bad aren't always clear, in the same way that an organism that eats unwanted weeds in your garden is seen as "good," until it starts eating the plants you're growing.) Certain gut microbes are never good and can be especially harmful, and you want to have enough "good" bacteria to keep them in check. Those invasive pathogens can cause temporary or chronic infection if they get through your gut lining (observed in inflammatory bowel disease—IBD). So the quality of the barrier between the microbiome and the tissue of your body is of critical importance.

The quality of the gut barrier depends on mitochondria in cells, which make up the gut lining. To protect against invasive pathogens, these cells create a two-tier defense system. First, to keep the dangerous microbes at a safe distance, these gut lining cells secrete a mucus layer, which serves as a buffer between them and the gut bacteria. This is the first line of defense. The second line of defense against pathogens is the gut lining cells themselves that serve as a wall against those dangerous bacteria, and all this makes clear why the gut lining cells need to be in top shape, with good supply of ATP—in other words, with good mitochondria.

There are things you can do to ensure the health of the gut barrier, and we think of this as a love triangle that involves you, your mitochondria, and some "good" bacteria that live in your gut. Here is one of several examples of that love triangle. To produce ATP, the mitochondria in the gut lining cells depend on a compound called butyrate. But our body is not very good at extracting butyrate directly from the food we eat; instead, we depend on some "good" bacteria that live in our colon that munch on dietary fiber to extract this compound, which allows mitochondria in the gut lining cells to produce ATP. Put succinctly: Your gut mitochondria need butyrate, which they get from "good" gut bacteria that depend on you to give them fiber from fruits, vegetables, whole grains, nuts, seeds, and legumes. Incidentally, butyrate is also found in butter (the word *butyrate* comes from the Greek word for butter). However, you would need to eat a lot of butter to get enough butyrate, and there are well-established reasons not to consume too much butter, so again, fiber it is.[3]

We usually hear about fiber as important for regular bowel movements, which is true, but only when you look at things at the cellular level do you start to recognize the tremendous role fiber plays in protecting us from the outside world.[4] Butyrate is just one example of a gift from the friendly bacteria in our gut that converts some things we can't digest into useful material. Two others are acetate and propionate, which, like butyrate, are short-chain fatty acids (SCFAs) that are created from resistant fibers (or complex carbohydrates) that cannot be digested, and travel all the way to the lower gut—to the colon. It is here that "good" bacteria convert them through fermentation into useful SCFAs. If you're still not convinced that dietary fiber is important, consider this: SCFAs can also offer some protection from colon cancer; in cases where these cells start to divide uncontrollably, some SCFAs have been shown to help by inducing programmed cell death (apoptosis) which, as mentioned, involves the mitochondria.[5]

The problem with fiber intake is that the average person in the United States consumes only around 15 grams of fiber per day, while the general recommended amount for adults is around 25 to 30 grams. As we just explained, insufficient fiber in your diet affects not only your bowel movement but also the important SCFAs that come from fiber. Inadequate fiber consumption is not limited to the States. Researchers from Massey University in New Zealand analyzed 243 studies and found that the highest dietary fiber consumers had a decrease of 15 to 30 percent in all-cause and cardiovascular-related mortality, stroke, type 2 diabetes, and colorectal and breast cancer relative to the lowest fiber consumers.[6]

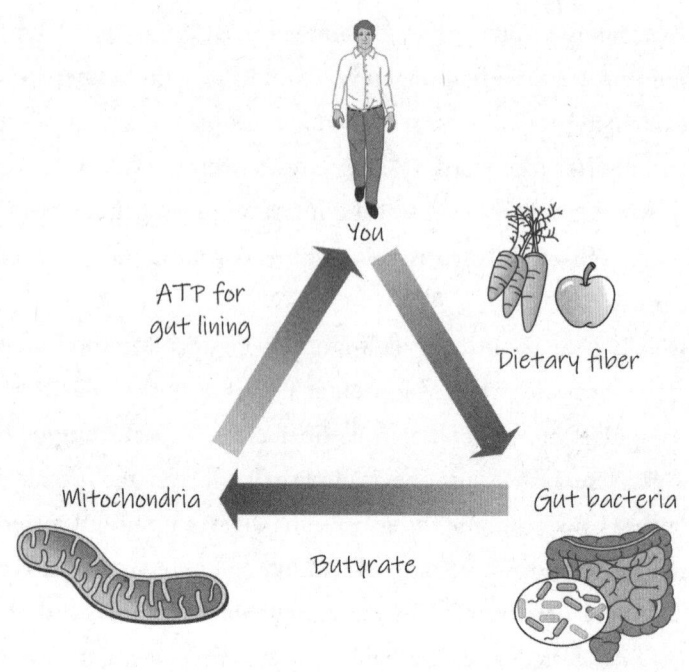

You

ATP for
gut lining

Dietary fiber

Mitochondria

Gut bacteria

Butyrate

One example of how eating fiber helps you. When you eat fiber, "good" bacteria in your colon convert it to butyrate, which allows the mitochondria in your gut lining to produce ATP. When the cells in the gut lining have the energy they need, they can protect you from inflammation and other problems.

How does this example relate to the commonly mentioned terms *prebiotics*, *postbiotics*, and *probiotics*? Let's quickly define them in our context. The word *prebiotics* refers to the ingredients you should give the bacteria in your gut (in the example above, fiber). *Postbiotics* refers to what the bacteria produce from what you gave them (in our example, butyrate). The term *probiotics* refers to "good" live bacteria that you add directly to your microbiome. This can be done through certain supplements containing "good" bacteria (but recall the cautionary notes regarding supplements' safety and lack of regulation that we made in the previous chapter) or through fermented foods that some call "nature's probiotics."

Examples of fermented foods include yogurt containing live and active cultures, kefir, kimchi, sauerkraut, and some pickles. More on fermented food later, but let's first look at three additional examples of love triangles.[7]

Pomegranate, Chocolate, and More

In 2021, dietitian, fashion model, and bestselling author Maye Musk posed for a promotional photograph wearing a white cashmere sweater embroidered with the phrase "I <3 my mitochondria." To understand what this was all about, and how it relates to the microbiome, think about pomegranates. Our gut bacteria can produce from pomegranates a compound called urolithin A, which has been shown to have wonderful effects on our mitochondria. Once formed in the gut, urolithin A is absorbed, resulting in the production of mitochondrial proteins (which means you can have mitochondria with newer, refreshed parts in them). Urolithin A also promotes mitophagy, which, as you may remember, helps get rid of damaged mitochondria. And it isn't only pomegranates; other foods that can be converted into urolithin A include strawberries, raspberries, walnuts, and pecans. Why did we mention the "I <3 my mitochondria" cashmere sweater here? There are indications that some people don't have urolithin A because they lack the right bacteria in their gut that produce it. Maye Musk participated in a promotion for a company that developed a molecule that unlocks a precise dose of urolithin A and thus helps the mitochondria, so this supplement can potentially be helpful for those without the right bacteria.[8]

You'll love the next example if you love dark chocolate: Chocolate has active ingredients that are very beneficial to the mitochondria. Some of them are not in the most active form, so while some are absorbed by the body directly, most of the good ingredients from dark chocolate need to be further metabolized by our microbiome into other compounds that have therapeutic benefits on the mitochondria and overall human health. The important ingredients to benefit the mitochondria are dark chocolate polyphenols, which have been shown to increase the number of mitochondria in tissues. Next time someone asks you why you eat dark chocolate, you can explain that you're doing it for your mitochondria (via your gut bacteria). Polyphenols can also be found in other foods, and some examples will be listed at the end of the chapter.

Our last example involves the protein-derived tryptophan, which can be found in a variety of foods (listed at the end of this chapter). When we eat chicken or chickpeas, for example, we get tryptophan, which is good for our gut bacteria. It is also the precursor for melatonin, which is usually known in the context of sleep.[9] But melatonin can also be found in the gut, and it is a powerful antioxidant—it promotes autophagy and mitophagy and attenuates gut dysbiosis (imbalance in the microbiome), intestinal barrier dysfunction, and inflammation. In addition to melatonin, when you eat foods that contain tryptophan, some of it is converted by gut bacteria into a compound called IPA. Sorry, not the beer; the IPA we're talking about stands for indole-3-propionic acid, which, in addition to its antioxidant effects, may improve mitochondrial functions by increasing the replacement of their components.[10] Again, mitochondria are instrumental in keeping your gut-lining cells healthy, so eating all these foods supports the quality of the barrier between the microbiome and the tissue of your body.

The more we read about the microbiome, the more questions we had about how to improve its health. Luckily, we didn't have to look far; the laboratory of Justin and Erica Sonnenburg, two leading experts in the field of microbiome, is a three-minute walk from Daria's lab, and we visited them on a spring morning to get their advice.

Keeping Your Microbiome Healthy

At Justin Sonnenburg's office, we were greeted by the couple and their family dog, Louis, a friendly white Havanese, named after Louis Pasteur, the French microbiologist who back in the nineteenth century saw the link between bacteria and disease. Louis (the dog, not the scientist) welcomed us with generous licks, and we reciprocated with back rubs and pats. During our visit, we learned that the Sonnenburgs are great believers in the importance of dietary fiber. They say that lack of fiber is driving inflammation, which is causing a wide range of Western diseases from heart diseases and type 2 diabetes to auto-immune diseases. They see the diet of highly processed foods and not enough fiber as a major factor that has led to microbiomes that are less diverse than before industrialization. A more diverse ecosystem (whether in the ocean, the jungle, or in our gut) is more resilient because each species has more options of what to eat, so the disappearance of one species doesn't cause the system to collapse.[11] The Sonnenburgs suspect that, in addition to poor diet, several other factors may have led to the deteriorated microbiome: Overuse of antibiotics, Cesarean sections, obsession with sanitation of our living environment, and limited physical contact with animals and soil. The discoveries of Louis Pasteur and his followers led to great advancements in sanitation that have saved millions of lives, and no one would want to go back to the days before running water and flush toilets. Yet many scientists hypothesize that drastically limiting any contact with soil and animals reduces our exposure to new "good" bacteria. For example, research suggests that families that own dogs have a more diverse microbiome than families that don't, and that babies who are exposed to dogs have a lower chance of developing asthma in the first six years of their lives.[12] (Thanks for the licks, Louis and friends!)

Over the years of studying the microbiome, the Sonnenburgs noticed that they were increasing their fiber consumption dramatically, while nobody else around them was. That was until they went to conferences with other microbiome researchers who were big on fiber just like them. So much so that at one conference, the staff at the food dining commons came up to them and asked what conference this was. "We said: 'We study the gut microbiome. Why?'" Justin Sonnenburg remembered. The dining staff said they can't keep the salad bar full. It seems that people who understand the importance of the microbiome adjust their diet, and since working on this chapter, we, too, have increased our fiber intake.

Feeding the gut microbiome is so important that it happens in babies' bodies even before they eat solid food. It turns out that breast milk contains ingredients called human milk oligosaccharides, or HMOs—complex carbohydrates that infants can't digest. Why does the mother's body produce something her baby can't use? The answer is that she nourishes the baby's microbiome; HMOs feed the "good" bacteria in the baby's gut.[13]

Good fences make good neighbors, and a study the Sonnenburgs ran in mice further explains why dietary fiber is so important to keep the "fence" between the dense community of microbes that live in the gut and the body. Recall that cells in the gut lining secrete a mucus layer that serves as the first line of defense against dangerous bacteria that live in the gut.[14] The experiment the Sonnenburgs ran was conducted on two groups of mice. One group was fed a high-fiber diet, and the other group was fed what would be the equivalent of a "doughnut and soft drinks" diet—lots of simple sugars and very few complex carbohydrates. They found that the mucus layer in the second group quickly became very thin, bringing the dangerous bacteria much closer to the tissue. In the absence of complex carbohydrates from the diet, the microbes in the gut relied on complex carbs that they found in the mucus. When the mucus layer gets too thin, dangerous bacteria may get through the gut lining. Indeed, they also found markers of inflammation in those mice.

What does it mean to us? If what happened in this mouse experiment is true in humans, too, it means that if you don't eat enough fiber, microbes may start munching on some complex carbs that are in the mucus, and as a result, that first line of defense might get thinner, letting dangerous bacteria way too close to the gut lining. This significantly increases the risk that pathogens will seep into the bloodstream, causing inflammation and infection.

To further appreciate the importance of having healthy mitochondria, consider this: When the mitochondria in the gut lining are dysfunctional, a fascinating (but dangerous) process can take place. The gut is largely an anaerobic environment, which means that there is very low oxygen in it.[15] The "good" bacteria that live in our gut thrive in this environment, and this includes the bacteria that make short-chain fatty acids (SCFAs) like butyrate. There are also "bad" bacteria that can operate with or without oxygen. (You can think of them as con artists who manage to thrive in every environment.) When mitochondria in the gut-lining cells work well, they consume oxygen from the gut to produce ATP, and so the gut stays relatively anaerobic (low oxygen), but when mitochondria in the gut-lining cells are dysfunctional, they do not use as much oxygen. As a result, there is much more oxygen inside the gut, which kills some "good" bacteria and allows those sleezy con artists to operate and form local abscesses or life-threatening infections. Yet another reason for why we want our mitochondria to be in top condition.

In addition to the benefits from dietary fiber, there is growing evidence for the health benefits of fermented food. The Sonnenburgs teamed up with Christopher Gardner at Stanford who uses randomized controlled trials to investigate the potential health benefits of various foods. Together, they started a small clinical trial to examine how increasing the consumption of dietary fiber and fermented food may lower inflammation and increase the diversity of the microbiome.[16] In this study, thirty-six healthy volunteers were randomly assigned to one of two groups. One group was asked to gradually increase their fiber consumption, while the other was asked to gradually increase the consumption of fermented food. Members of the fiber group had a goal of adding at least 20 grams of fiber per day by consuming more foods like legumes, seeds, whole grains, nuts, vegetables, and fruits. Members of the fermented-food group had the goal of reaching at least six servings a day of fermented foods, and they were encouraged to include a variety of foods that contained live and active microbes—like fermented dairy products, fermented vegetables, and fermented nonalcoholic drinks. Importantly, all participants were asked to achieve the increase from food and not from fiber supplements or probiotic supplements; other than this, participants were not restricted in what they could eat or drink. During the six weeks that followed the four-week ramp-up stage, participants were instructed to maintain the high level of consumption of fiber or fermented food according to their group assignment.

While this was a small study, it has four main takeaways. The first one is that it can be done. With a few simple guidelines, people shifted to a more microbiome-friendly diet. During the study, the high-fiber diet group ended up eating about 40 grams of fiber per day, and those in the fermented-food group increased their consumption to about six servings of fermented food per day. Second, the study showed that a fermented-food diet increased the microbiome's diversity and decreased signs of inflammation. (Again, a more diverse microbiome is more resilient; inflammation is a major driver of chronic diseases.) Third, contrary to the researchers' prediction, the high-fiber diet did not increase the microbiome diversity, but their microbiome shifted toward one that was better able to degrade complex carbohydrates. Fourth, members of this high-fiber group enjoyed other health benefits; for example, they consumed less processed food, possibly because they felt full earlier.

The researchers also expected a decrease in inflammation in the high-fiber group, and this indeed happened, but not among participants who started the study with a deteriorated microbiome with low diversity. For some of those latter participants, markers of inflammation went up, but the Sonnenburgs believe that this is likely due to the short-term intervention; the deteriorated microbiomes did not have enough time to recover the diversity needed to benefit. They added that it is likely that a high-fiber diet over a longer period of time would be widely beneficial. Overall, this study demonstrated mostly the importance of fermented food in decreasing inflammation and increasing microbiome diversity. But of course, in real life, it's not either/or, so the bottom line is that many of us can benefit from increasing the consumption of both fiber and fermented foods.

More Touchpoints of the Two Ancient Tribes?

As we pointed out, only in recent years have researchers begun to study the interaction between mitochondria and the microbiome. So the last word about their intersection has not been spoken. Here are just three more areas of potential interface.

One possible touchpoint is emotional stress. Later in the book, we'll discuss the link between emotional stress and mitochondria, and we've all experienced the link between the microbiome and stress firsthand (or "first gut") before an important exam or presentation. There is solid evidence that reducing chronic psychological stress is good for both our microbiome and our mitochondria. Relatedly, there is evidence for links of these two tribes with our brain. On the microbiome side, there is the "Gut-Brain Axis," a two-way communication channel between the gut microbiome and the central nervous system. Meanwhile, on the mitochondria side, researchers have demonstrated a close link between mitochondrial dysfunctions and psychiatric disorders as well as with neurodegenerative diseases, some of which we discussed. Although mitochondria and gut microbiome research are done by different groups of scientists, some researchers are starting to draw attention to the crosstalk between mitochondria and gut bacteria in the context of the brain, which may lead to new insights on how to address neurodegenerative diseases and psychiatric disorders.[17]

A second possible touchpoint is around immune response. For example, both mitochondria and gut bacteria generate similar short peptides to recruit immune cells to sites of infection in the body, suggesting possible coordinated signaling between them. The fact that these two communities "speak the same language" of similar compounds may suggest that there's a communication that we don't yet fully understand.

A third intriguing area suggesting a link between mitochondria and the microbiome is pain. Hyperalgesia (heightened sensitivity to pain) is associated with increased innate immune response, where mitochondria play a key role. Some studies show that mitochondrial dysfunctions contribute to chronic pain, and that reestablishing mitochondrial functions may decrease chronic pain. For example, Daria's lab found that activating enzymes that remove aldehydes inside the mitochondria greatly reduces pain in a mouse model of inflammatory pain.[18] At the same time, the gut microbiome generates a number of neurotransmitters that may help in pain management.[19] These include serotonin (a calming neurotransmitter) and dopamine (a reward-system neurotransmitter) to name two. Research indicates that gut bacteria communicate with the brain.[20] Indeed, a 2021 small study in Ireland of twenty patients after limb surgery demonstrated that consumption of pain medication seven days after surgery was inversely correlated with the diversity and composition of their gut microbiome. Those with better microbiomes were able to tolerate pain more easily.[21]

Fine-tuning Your Gut Microbiome

Here are things to consider when trying to improve your gut health. The microbiome consists of hundreds of different types of bacteria and other organisms, with numerous possible combinations that can get out of balance (dysbiosis) with serious consequences. Experts therefore recommend changing your diet gradually and gently and tuning in to how your body reacts to these changes.[22] The microbiome can respond quickly to changes in your diet, but it can easily revert to its previous composition, so adhering to a plan that works is recommended. Keep in mind that the microbiome is unique to each individual, so what works for some people won't necessarily work for others.

Adjust your dietary fiber intake, if necessary. Remember that the mitochondria in your gut lining produce ATP from compounds such as butyrate that they get from bacteria that feed on fiber. Your end of the deal is to provide that fiber. While working on this chapter, we printed a couple of charts of high-fiber foods we had downloaded from sites such as the Mayo Clinic website and left them on our kitchen counter.[23] We've always eaten a lot of fruits, vegetables, nuts, and seeds, but when we started measuring our intake, we realized that we could do even better in reaching the general recommended amount of daily dietary fiber for adults, around 25 to 30 grams. For example, we learned that a salad with 3 cups of shredded lettuce, a cucumber, and an average-sized tomato will give us only 5.5 grams of dietary fiber, but that if we add half an avocado, we get it up to 10.5 grams. We also quickly realized that we could increase our fiber consumption dramatically by eating legumes like lentils, beans, and chickpeas more often. Since different fibers do different things, relying on a variety of sources of dietary fibers is a good idea.[24]

Consider trying more fermented food. We've known for a long time that fermented foods are good for us, so there's always been yogurt and kefir in our fridge, but we didn't eat them every day, which we now try to do. When shopping, be aware that the label on most fermented foods indicates that it contains live and active cultures, and because it contains living organisms, it is stored in the refrigerated area of the store. We're also trying to expand the variety of fermented food. (Emanuel went back to his German roots and increased his sauerkraut consumption.) It's important to note that if you cook fermented food like sauerkraut, you kill the live bacteria, so you won't receive any benefits, and cucumbers that have been brined in vinegar are not fermented food.

Minimize ultra-processed food. The removal of fiber in foods such as white bread or certain snacks is part of ultra-processing, where instead of fibers you get lots of glucose, fructose, and sucrose, giving you the sweet hit and calories without the support of natural fiber.

Be mindful of side effects of antibiotics. Antibiotics can be very helpful and even lifesaving, but they can also cause disruption in the microbiome balance because, in addition to killing bacteria that cause infections, antibiotics also kill some "good" gut bacteria.[25] Many experts, therefore, agree that one should avoid *unnecessary* courses of antibiotics. In certain cases, antibiotics are necessary, and when weighing the pros and cons, they will still be prescribed. Is it a good idea to use probiotics supplementation to help your "good" bacteria during or after antibiotics? We'll discuss this question and probiotics, in general, next.

The jury is still out on probiotics. Probiotics supplements contain "good" bacteria that may reinforce your microbiome. Many physicians recommend their use when taking antibiotics since they have been shown to prevent diarrhea that is associated with antibiotics. But some studies show that probiotics can sometimes disrupt the balance in your microbiome.[26] Meta-analysis of probiotics use conducted in 2021 found somewhat more encouraging results for patients with metabolic syndrome.[27] (Meta-analysis draws conclusions based on data from multiple independent studies.) Since probiotic supplements (like other supplements) are not regulated, many experts are hesitant about them and recommend trying nature's probiotics—fermented food which contains live and active bacteria as well as fiber. The bottom line about probiotics is that it is too early to tell. Yes, probiotics can be helpful sometimes for some people and have been shown to be beneficial in some cases of irritable bowel syndrome and inflammatory bowel disease. However, this is a young area of research, and the answers may be highly individualized.

If things aren't right, seek professional help. If you regularly experience discomfort, upset stomach, bloating, pain, constipation, or diarrhea, seek professional help. The above is general advice that experts believe can help most people, but there are treatments not covered here for people with irritable bowel syndrome, celiac disease, and other gastrointestinal disorders.

Examples of Foods Beneficial to Your Gut

High-Fiber Foods	Whole grains (e.g., oatmeal, brown rice, whole wheat), legumes (e.g., beans, lentils, chickpeas), fruits (e.g., apples, apricots, bananas, pears, raspberries), vegetables (e.g., carrots, artichokes, asparagus, broccoli, garlic, onions, green peas, leafy greens), nuts, and seeds.
Foods That Can Be Converted into Urolithin A	Pomegranates, strawberries, raspberries, walnuts, and pecans can be converted by our gut bacteria into urolithin A, which helps mitochondria.
Foods Containing Polyphenols	Numerous foods contain polyphenols, and here are some examples (in addition to dark chocolate): apples, berries, broccoli, carrots, cumin, flaxseeds, ginger, green tea, oats, olives, olive oil, onions, red cabbage, sesame seeds, spinach, turmeric, and whole grains.
Foods Containing Tryptophan	Oats, dried dates, milk, yogurt, cottage cheese, red meat, eggs, fish, poultry, sesame, chickpeas, almonds, sunflower seeds, pumpkin seeds, peanuts, and more.

• • •

New tools are emerging to test the composition of your microbiome. These at-home testing kits can give you an idea regarding its diversity. Since each microbiome is a unique ecosystem that evolved over one's lifetime, translating this knowledge to specific dietary recommendations is challenging, although periodic monitoring might help you assess how changes in your diet may affect your microbiome. Perhaps one day, we'll be able to manage our gastrointestinal tract with the precision of a well-run chemical plant, but until then, we need to rely on our "gut feeling," so to speak. How are you feeling after adding fermented food to your diet? Have things changed since you started eating more fiber? To see if your stool looks normal, the Bristol stool chart (that can be easily found online) is a widely used instrument. Essentially, your stool shouldn't be too hard or too soft, and it should pass through easily. Regular bowel movements are important, too, and they can be regulated by consuming dietary fiber.

Helping these two tribes live in peace together is important. Your microbiome can then help your mitochondria with essential nutrients like B vitamins. For their part, your mitochondria can prevent harmful organisms from seeping into your blood. Your end of the deal is to feed your gut bacteria with what they need. It is also worth noting that your microbiome is influenced not only by your diet, but also by things like your sleep pattern and psychological stress. Scientists also point out to links between exercise, mitochondria, and the microbiome. We mention all this just as a reminder that when it comes to interventions, it's hardly ever one thing, and that working on multiple fronts is a good idea. In the next chapter, we show that it's not all about hard work. Your mitochondria also need a break.

CHAPTER 9

Give Them a Break

Recall that the life machines create pollution as you go through your daily activities, and that this pollution can damage them. But what if scientists came up with a treatment to repair at least some of this damage? And if they did, how much would you pay for such treatment? The good news is that such a treatment exists, and it won't cost you a dime. Just relax and get a good night's sleep.

During the day, our mitochondria focus on energy transfer, and during the night, they focus on recovery. We've seen in previous chapters that this recovery is done through autophagy, mitophagy, and mitochondrial dynamics (fission and fusion). To refresh your memory of how these processes work, damaged bits and pieces are isolated in one end of a mitochondrion and that section is collected by autophagosomes (the "garbage trucks") and brought to lysosomes (the "recycling centers") to be digested to their building blocks, which are later used to build new cell components. Mitochondria do this for us every night. Our end of the deal is to let them do it by sleeping and giving them enough time, not overwhelming them with additional jobs, and not confusing them or distracting them.

In writing this chapter, we talked to Erin Flynn-Evans, who is helping a group of people facing an especially challenging sleep environment. These folks sleep standing up in sleeping bags attached to the wall; sometimes upside down. They are always surrounded by noisy machines, and sometimes they need to sleep in a very crowded space. If that's not enough, they experience sixteen sunrises in twenty-four hours. We're talking, of course, about astronauts.

175

Flynn-Evans is the director of the Fatigue Countermeasures Laboratory at NASA Ames Research Center, and we thought that if she shows us that these people can be helped, it means that the rest of us here on Earth can maybe be helped, too. The list of the challenges astronauts face is endless. For an optimal sleep environment, noise should be kept under 35 decibels, which is roughly around the sound of whispering, but the base-level noise in the International Space Station is around 80 decibels, which is like having a vacuum cleaner running continuously in your bedroom. Further, while a cool sleep environment is usually recommended, it sometimes gets so cold that one astronaut reported sleeping with their laptops by their feet to keep their toes warm. And because of microgravity, sometimes astronauts are awoken by their own hands accidentally hitting their face. Is it possible to help them sleep better with all these distractions? The answer is yes. We'll discuss measures NASA is taking in this area and what we can do here on Earth to improve our sleep environment, too.[1] To understand these measures, let's explain some basics of sleep and how mitochondria are involved.

What Makes You Tired? What Makes You Sleep?

Sleep is driven by two processes: the homeostatic and the circadian. The homeostatic process is simply the buildup of the need to sleep. You can think of it as an hourglass that measures how tired you are, except this hourglass is not filled with sand, but with a compound called adenosine. If the name sounds familiar, it's because we mentioned it when we introduced adenosine triphosphate (ATP). Recall that the brain uses about 20 percent of the ATP in our body, and its consumption results in adenosine accumulation in the brain. This accumulated adenosine provides one of the signals to your body that you need to sleep. When a relatively small amount of adenosine accumulates in the extracellular space in your brain, the message is: "I'm tired." When a lot of adenosine accumulates there, the message is more of a scream: "I'm damn tired, and I need to sleep right now!" It's a beautiful system: the more your brain works, the more ATP is consumed, and a by-product of ATP consumption is being used to monitor your brain's need to rest for a major cleanup.

And this process is not limited to the brain; other cells release adenosine as they "work." For example, the more your muscles work, the more adenosine accumulates between the muscle cells, and this accumulation also tells your brain that you need to sleep. All this becomes important to understanding two substances—caffeine and alcohol. When you drink a cup of coffee or a glass of wine, you are really managing adenosine levels in your system. Caffeine inhibits the response to adenosine that accumulated in your brain, so it makes you *feel* less tired. Unfortunately, this may give you the impression that you're fine, even if you're sleep-deprived. Alcohol increases the accumulation of adenosine by inhibiting its uptake, which makes you fall asleep faster. But be wary, we'll see later in this chapter that the use of these substances isn't without problems when it comes to getting a good night's sleep.[2]

The second process that is important to sleep is the circadian system. Right above the point in our brain where the two optic nerves cross, there's a little region called SCN (full name suprachiasmatic nucleus), which synchronizes our circadian rhythm. Its location is not a coincidence because it uses valuable information that comes from the eyes through the optic nerves—information about light. Specifically, there are special light-detecting cells in our eyes, and when they notify the SCN that there's light outside, it sends both neural signals and hormones that it is daytime. Conversely, when it is dark, the pineal gland, which is in your brain as well, sends melatonin to your body with a simple message: It's time to sleep. Importantly, melatonin is produced in the pineal gland only when it is dark. What does it mean in terms of getting a good night's sleep? We'll elaborate on this later, but it's not hard to guess: When it comes to sleep, darkness is your friend. There are individual differences in people's circadian rhythms, of course. Some of us are early birds, others are night owls, while others can be in the middle, but we all do respond to sunlight.

A simplistic view of the circadian system goes as follows: Our brain manages our schedule based on sunlight, in the same way that our phone tells us when the next meeting starts. Yet even if you're the busiest person with the craziest schedule, the task of managing your body is much more complex than managing your daily meetings and requires a more sophisticated system than just one clock. In fact, every cell in your body has a clock that consists of several proteins that are produced and broken down in a twenty-four-hour cycle. (We won't elaborate on these proteins here, but for those who are interested, some of them are Bmal1, Clock, Per, and Cry.) Depending on the organ or tissues, these cellular clocks work a bit differently, so scientists in the field talk about several peripheral clocks in different organs such as the liver, the kidney, the lungs, and your skeletal muscles. A key point is that those clocks aren't only stimulated by light. Your food and your activity are two examples of other inputs, and understanding that our body uses different cues regarding time makes it clear why the SCN in our brain is so important. With so many clocks around, it's essential to have a master clock that synchronizes all these tiny clocks with the day-night cycle. Without the SCN, our body would be like a chaotic clock shop where every clock shows a different time. Just imagine the cacophony of constant beeping of digital clocks, chirping from cuckoo clocks, and chiming of grandfather clocks. Luckily, we have a beautifully coordinated system, unless we screw things up.[3]

As metabolic hubs, mitochondria are involved in building some of the compounds we have mentioned: As part of the homeostatic system, ATP that the brain consumes is degraded to adenosine, which accumulates outside neurons to drive our need to sleep. In the circadian system, mitochondria are involved in the production of melatonin, which is the hormone that makes us sleepy. And another important hormone that we haven't discussed—cortisol, a so-called stress hormone—is partially synthesized in mitochondria and secreted in the morning to help us start our daily activities.

What Happens When You Raid Your Fridge at Night?

It's eleven p.m., you're watching reruns of your favorite show, and the leftover chocolate cake in the fridge keeps calling your name. What happens if you give in and eat it? Before you take the first bite, your master clock is in sync with the peripheral clocks; the clocks in your liver, fat tissue, and skeletal muscles are working in harmony, preparing your body for sleep. But once you start eating that cake, the clocks in some tissues get a strong signal that changes their time. Now the clocks in your liver and fat tissue operate not based on the message that came from the master clock (which is affected by the sun), but on the message they just got from that chocolate cake: "Time to work!" Why is this bad? Because your cell clock regulates what's happening in your mitochondria, and now the mitochondria in your liver are getting conflicting messages.

Earlier in the evening, there was an important switch in your mitochondria while they were preparing for the night. They switched from metabolizing carbohydrates during the day to metabolizing stored fatty acids during the night. This switch happens because our body needs energy throughout the night, but glucose can be stored in the body (in the form of glycogen) only for a few hours. In contrast, fat is stored in our body indefinitely. By using fat at night, mitochondria have the raw materials they need to produce ATP for all their recovery activities: triggering the removal of damaged parts through mitophagy and autophagy, cleaning up "soot" and "rust," in addition to the other activities that are still going on, like generating building blocks.

As humans have evolved, our bodies have been built to deal with periods of feeding and periods of starvation, during which the body uses fat. Nights (our normal sleeping time) are periods of no calorie intake (kind of starvation), and the circadian clock evolved to prepare for the night by turning on expression of fat-metabolizing enzymes. "Our liver is not the same liver during the day and during the night in terms of its composition of proteins and enzymes," Orian Shirihai from UCLA, a long-time researcher in the field, explained to us. Your liver is ready to deal with fats at night, and sending it an injection of sugar in the middle of the night is like making a reservation for fifty single rooms in a hotel and then having fifty families show up, each with a few kids. Just think of the poor receptionist when they all show up at eleven p.m. in four charter buses. When you eat at night, you disrupt your circadian rhythm, and you confuse your mitochondria's metabolic switch, interrupting their self-cleaning, all of which can harm your liver, as we discuss next.[4]

Eating fats at night isn't good either, because your mitochondria should be metabolizing the fats you *already* have in the system and not be overwhelmed with digestion, distribution, and storing new food. The benefit of restricting eating to daytime hours was demonstrated in a 2014 study from the Weizmann Institute in Israel. This study was conducted in mice (who are active at night) and showed that if mice eat only at night when they are active, rather than throughout the day and night, they will eat the same number of calories, but their liver lipid levels will be 50 percent lower.[5] Mice are not humans, but this study suggests that avoiding food late at night may be good for the human liver, too. This is especially important since accumulating fat in your liver is detrimental to your health: nonalcoholic fatty liver disease, NAFLD (and its inflamed cousin NASH—nonalcoholic steatohepatitis) are the most common liver disorders in industrialized countries. If these conditions are not kept in check, they can lead to cirrhosis, scarring of the liver, and in extreme cases liver failure.[6]

If you're good about avoiding food at night, you're doing yourself a great favor, but the link between sleep and proper metabolism doesn't end with staying away from the fridge at night. Many studies examine the link between short sleep and obesity. For example, out of twenty-eight longitudinal studies in this area, twenty-two reported that short sleep (generally less than six hours per night) is associated with weight gain or development of obesity and/or increased fat mass.[7] There is also a strong association between sleep problems and metabolic inflexibility, which, as you may remember, is a condition in which mitochondria lose the ability to switch between metabolizing carbohydrates and metabolizing fatty acids. Recall also that metabolic inflexibility is associated with insulin resistance, which can lead to type 2 diabetes. It isn't clear which comes first: Does metabolic inflexibility lead to sleep problems? Or do sleep problems lead to metabolic inflexibility? But we know that the two tend to go hand in hand.[8]

Many questions regarding cause and effect in this field are still open, but there is strong evidence that mitochondrial dysfunctions are associated with sleep problems. When this happens, all the beautiful (and critically important) processes that we described earlier in the book, can be disrupted: the regeneration of mitochondrial components, the fusion/fission dance (mitochondrial dynamics), and the cleanup process (mitophagy and autophagy), are not working correctly. And this doesn't affect your body only during the night. While this chapter focuses on sleep, circadian rhythms affect our activities every minute. Everything in our body runs on a twenty-four-hour schedule, and this may be related to the timing and the content of our meals throughout the day.[9]

In addition to food, your sleep can be affected by alcohol, and it isn't a positive effect.[10] Although it helps you fall asleep faster because it increases the accumulation of adenosine (which builds up your need to sleep), alcohol negatively affects your sleep in three ways. First, mitochondria in your liver that should be helping with the nightly cleanup of your body and fatty acid metabolism are now busy metabolizing alcohol instead. And the metabolite of alcohol is an aldehyde, acetaldehyde, which is a reactive compound that adds to what we called "rust" or "soot" in the cells; like rust or soot accumulating in an engine, aldehydes stuck on cell components prevent them from operating properly (more on this in chapter 11). Second, alcohol disrupts histamines in the brain, which regulate the sleep-wake cycle. Third, research shows that your sleep quality suffers when you drink alcohol because there is less time for the type of sleep that is considered restorative—rapid eye movement (REM) sleep. Yet alcohol is not the only cause of poor sleep quality, which we discuss next.

Poor Sleep Quality: Sleep Misalignment

We've all slept on flights, so at first glance, there was nothing special about the two men dozing next to each other on the flight that left Kendari, Indonesia, to the capital, Jakarta, in January 2024. However, about an hour and fifty minutes after takeoff, air traffic control lost contact with the plane that veered off its designated path. The problem was that the two men sleeping were the pilots, who, according to a preliminary report from Indonesia's National Transportation Safety Committee, were both asleep at the same time for about half an hour, and luckily woke up in time to land the plane safely. Getting enough sleep is important, and based on the report, it doesn't seem like the pilots got enough sleep before the flight. How much sleep is enough? This depends on your age. For adults between twenty-six and sixty-four, the recommended range is seven to nine hours. Of course, actual needs can vary from person to person. In addition, older people are believed to need a little less sleep, while babies, children, and teenagers need more, in part because growth hormones are also active during sleep.

But as the case of the pilots illustrates, it isn't only about how many hours you sleep. The duration of one's sleep is only part of the overall sleep quality, and according to the report, one of the pilots had one-month-old twins at home, and before the flight, he had to wake up several times to help his wife take care of the babies. Fragmented sleep can mean that you don't have enough REM sleep which, again, is considered restorative and also helps with learning and memory. A lot of cleaning is believed to be happening during REM sleep when we dream. In 2024, a study in mice by Jonathan Kipnis of Washington University in St. Louis, suggested that the "waste" accumulating in the brain due to neuronal activity is removed mainly at night, when we are actively dreaming; the synchronized brain waves during REM create rhythmic motion in the brain's fluid that is required for draining the waste out.[11]

Both sleep deprivation and poor sleep quality such as fragmented sleep have been linked to dysfunctional mitochondria, at least in mice, which may explain that sluggish feeling and why you want to crawl back into bed after pulling an all-nighter to finish an important report or staying up late to watch reruns of your favorite show. We must give mitochondria enough quality time to repair themselves at night.

Circadian misalignment (a mismatch between your internal clock and the time on your cell phone) is another critical issue when it comes to sleep. Unsurprisingly, it is one of the biggest ones in the International Space Station, which goes around the Earth every ninety minutes. Astronauts thus live in approximately forty-five-minute light and forty-five-minute dark cycles. An example of misalignment that is more familiar to most of us is jet lag. In his book *Pattern Recognition*, William Gibson writes about a young marketing consultant who travels between Tokyo, Moscow, and London. Crushed by jet lag in London, she feels like her soul was left behind at her last destination. "Souls can't move that quickly, and are left behind, and must be awaited, upon arrival, like lost luggage," Gibson wrote.[12]

We'll never forget a trip to Paris with our four kids. We arrived in the morning and decided to take a short nap to refresh before we explored the city. After an hour, when we decided to wake up the kids to start our exploration, we discovered one of the basic rules of the universe: that two tired adults cannot wake up four jet-lagged kids. We managed to wake up two of them and drag them to wait by the door, and went to wake up the other two, only to discover that the first two were back in their beds, face down. It's as if their bodies were in Paris while their little souls were still in California. Incidentally, speaking of Paris and souls: the French philosopher René Descartes believed that the pineal gland (which sends melatonin when it's dark) is the body part that is most closely associated with the human soul. While this is something that will be hard to prove, "amazingly, he considered that the gland itself was controlled by light," the late Josephine Arendt, a pioneer in melatonin research, said in a lecture a few years ago.[13] On the practical side of dealing with jet lag, when traveling east, it is a good idea to take a walk in the sun when you arrive in a city in a different time zone. It's a good idea even if you don't travel. Many experts recommend exposing yourself to sunlight in the morning to signal to the SCN that the day has started. This way, you help your body gradually adjust to seasonal changes.

Consistent misalignment and chronic sleep loss are associated with certain diseases, but we want to be careful here. Some articles about sleep freak people out by listing all the diseases associated with sleep problems. While sleep disorders can have a negative impact on your health, it is important to understand several things. First, we can't lump together all sleep problems. Stating that sleep disorders have been linked to chronic diseases and depression isn't very helpful because the situation is more granular; there are several different sleep disorders, at different levels of severity. Second, correlation is not causation, and it's possible that other variables play a role here. One strong candidate is emotional stress. "I have a sneaking suspicion that many of the things that we ascribe to sleep loss, have nothing to do with sleep, but they have much more to do with stress," sleep researcher Jamie Zeitzer from Stanford told us. Third, keep in mind that sleep is just one element in your overall wellness, and some behaviors can compensate for others, at least to some extent. Fourth, and most important, the good news is that tweaking certain habits is likely to improve your sleep, and many people who seek professional help improve their sleep. Still, while we don't want to scare people, we feel that it would be irresponsible on our part to ignore the research on this topic. Several reliable reviews demonstrate an association between certain sleep problems and certain diseases, including heart and neurodegenerative diseases.[14] So this cannot be taken lightly, and should point out the importance of good sleep, and hopefully motivate you to adjust your habits and seek professional help if you still have a problem. Covering these diseases is beyond the scope of this book, but here are several points that relate to mitochondria.

In an intriguing 2015 study, Danish and American researchers identified fifteen pairs of identical twins where one of the twins regularly sleeps between seven and nine hours (the recommended duration), and the other twin sleeps under seven hours.[15] Analyzing the mtDNA of the thirty participants showed that the twins who slept less had fewer mitochondria, which, as mentioned, is not the desired situation. Additional studies show that getting enough sleep is associated with healthier mitochondria and, thus, a healthier body. For example, a study conducted in Pakistan on eighty subjects found that those with good sleep quality had a higher mtDNA count number than those with poor sleep quality.[16] Researchers from Korea also confirmed this. After analyzing the sleep data of 238 middle-aged people, they concluded that poor sleep quality "is associated with reduced mtDNA count number, suggesting a potential biological mechanism whereby poor sleep quality, specifically long sleep latency [the time it takes a person to fall asleep], accelerates cellular aging and impairs health through mitochondrial dysfunction."[17]

If all this sounds too gloomy, here's another study that suggests that there's something we may be able to do to alleviate the effect of poor sleep. In this 2021 study from Victoria University in Melbourne, Australia, a group of eight healthy individuals were subject to sleep restriction (four hours a night) for five nights. By now you can guess that the mitochondria of this group suffered; there was a reduction in mitochondrial respiratory function. However, there was another group in this study: This group of eight healthy individuals was subjected to the same sleep restriction (four hours a night) for five nights, but they also did high-intensity interval exercises on each of these five days. The mitochondria of members of this group were as healthy as the mitochondria of a third group of eight healthy individuals, who slept eight hours a night for five nights. The point of this study is that the damage from poor sleep can be mitigated at least to some extent by intense exercise.[18] It does *not* mean that you can settle for four hours of sleep per night and compensate for that with exercise. Sleeping four hours a night on a regular basis is a bad idea for the vast majority of people; we're only trying to point out that certain behaviors can compensate for others from time to time. Having a fun evening with family and friends that lifts your spirit and reduces stress may be worth it, even if it cuts into your sleep time.

Developing Better Sleep Habits

What can we do to sleep better? To answer this, let's go back to astronauts. Over time, NASA has implemented several changes to create a better sleep environment, and despite the huge challenges of sleeping in space, as a group, astronauts are sleeping better than they have in the past. "Especially compared to history, it's a huge improvement," Erin Flynn-Evans from NASA's Ames Research Center told us. If they can do it in space, most of us can do it here on Earth. Based on these findings and on studies from other experts in the field, here are a few guidelines that may help you, too.

Give sleep a chance, and keep regular hours. An obvious, but often overlooked, rule of good sleep is to allow yourself enough time to get your appropriate number of hours. Regardless of whether you're an early bird or a night owl, if you binge-watch your favorite show until one a.m. and your alarm clock is set for five thirty a.m., there is no way that you will get enough sleep. Again, most adults between twenty-six and sixty-four need seven to nine hours. Babies, children, and teenagers need more sleep, and people over the age of sixty-five are believed to need a little less. There used to be no rules regarding the minimum number of sleep hours for astronauts, but today, astronauts are given a window of eight and a half hours for sleep. And not just any random block of eight and a half hours. Keeping a regular sleep window is important, too. In the past, if an astronaut was needed to replace a part or perform another maintenance task, they had to do it regardless of their usual sleep schedule. Today, keeping consistency for the sleep/wake timing is a big part of planning. This was one of several changes that were initiated all at the same time, which makes it hard to disentangle its effect from other interventions, but Flynn-Evans believes that this is one area that significantly contributed to the improvement. Here on Earth, the importance of a consistent bedtime is well documented. In 2023, a panel of experts reached a consensus on the issue of sleep regularity: "Consistency of sleep onset and offset timing is important for health, safety, and performance," the experts wrote.[19] In short: Try to go to sleep and wake up at the same time every day. "Bedtimes aren't just for toddlers," Flynn-Evans says, and recommends picking a bedtime and sticking to it as much as you can.

Be mindful of light and darkness. Light is a powerful force in regulating our circadian rhythm. One low-tech solution at NASA was to install window shades in space vehicles to prevent light exposure at the wrong hours. "Seems quite obvious," says Flynn-Evans, "but we didn't have window shades for a long time." A more sophisticated solution is the special lights installed on the space station that shift from a blue wavelength light in the morning to more of a red wavelength light in the evening. The light-detecting cells in our eyes that notify the SCN that there's light outside are particularly good at detecting blue light, the light they get from the sun. This is perfect in the morning, but blue light in the evening sends our body the wrong message. What does it mean in terms of getting a good night's sleep here on Earth? Sleep experts point out four things:

1. Being exposed to sunlight first thing in the morning will send your SCN a two-word message: Good morning! Because you spent the previous few hours in darkness, your system is more sensitive to light in the morning, so even a relatively small amount of light can be effective. (Remember, never look directly at the sun because it can permanently damage your retina.)

2. Spending time outside during the daytime helps, too. This is true even if it's cloudy. You're getting much more light outdoors than indoors.[20] Being outside makes the master clock more robust and helps it synchronize not only with the outside day, but also with peripheral clocks in your body.

3. Be mindful of blue light–emitting screens. Some experts recommend avoiding screens at least an hour before your usual bedtime. It seems that the effect of TV screens at a distance is not as detrimental to your sleep as screens that are close to your face. If you were exposed to a lot of natural light outside during the day, you may be less affected by the light from the screens. Kids are more sensitive to light, so they are more likely to be affected.[21]

4. Many experts agree that you should sleep in a cool and totally dark room and not turn on the light if you wake up in the middle of the night, especially not any equipment that emits blue light, like your phone or tablet. While the evidence for this recommendation is strongest for children, the body of evidence continues to evolve for other age groups. "The influence of light never ceases to amaze me in that every year it seems we learn something new about how powerful light is and how [even] little light exposure is impactful," Flynn-Evans says.

Avoid caffeine later in the day. When it comes to sleep, caffeine should be used carefully. NASA astronauts use caffeine to help them be more alert when they wake up, but they are advised against usage later in their "day" because caffeine stays in the system for a long time and can increase the time to fall asleep and make the sleep less restful. For some people, the effect of caffeine can be shorter (often because their liver eliminates caffeine faster), but our ability to objectively assess our sleep quality is limited, so caffeine intake in the afternoon is not recommended. As mentioned, caffeine inhibits adenosine's effect, and thus makes you feel less tired. This can create the illusion that you're getting enough sleep, when in fact you're not fully benefiting from the restorative effect of a good night's sleep. "The idea that you can use caffeine to replace sleep is definitely not true," says Jamie Zeitzer from Stanford.

Manage your eating. Staying away from the kitchen at night can be greatly beneficial to your mitochondria, and avoiding food a couple of hours before bedtime is a good idea, too. Food at night can do two things to your mitochondria. First, it can overwhelm them because they are busy with cleanup. Giving them the additional task of metabolizing new food adds to their workload at the expense of their main task. Second, it can confuse your mitochondria because food tells your liver that it is daytime.

Limit your alcohol intake. Alcohol may help you fall asleep faster because it increases the accumulation of adenosine, but it won't give you the quality sleep you need. As mentioned, research indicates that sleep quality suffers after drinking alcohol because there is less time for REM sleep, which is considered restorative. As we'll explain in chapter 11, alcohol metabolism also generates aldehydes that contribute to mitochondrial damage.

Exercise. Recall that when adenosine accumulates between your muscle cells, this tells your brain that you need to sleep. The following 1998 study from the University of Copenhagen in Denmark demonstrated how exercise affects the accumulation of adenosine. When seven healthy male volunteers exercised lightly (fifteen minutes of knee extensions at a light work rate) the adenosine level between their muscle cells increased fivefold, and when the workload of the exercise was increased, adenosine levels rose up to tenfold higher than the resting levels.[22] We all have had the experience of "sleeping like a baby" after a strenuous hike or a grueling gym session, and indeed, meta-analysis identified twenty-nine studies showing the positive effect of exercise on the duration or quality of sleep. However, the same analysis also included four studies that found no effect and one that identified a negative effect of exercise on sleep.[23] Overall, exercise helps with sleep duration and/or quality by increasing the levels of adenosine, which inhibits the neuronal network that induces wakefulness. As for the question of when it is best to exercise—although some researchers argue for the benefit of exercising in the morning, the evidence on this seems to be mixed. So at least until there is strong evidence, exercise when it works best for you.

Relax. We learned the hard way that difficult conversations before bedtime don't lead to a good night's sleep, so unless it's urgent, we don't start such conversations in the evening. It's also a good idea to avoid other potential anxiety-invoking things such as email or social media before bedtime. (We try to be good about this; Emanuel is still struggling.) There is clear evidence that stress interferes with sleep, and vice versa: bad sleep leads to stress. Luckily, there are ways to reduce stress before bedtime. For some people, taking a warm shower does the trick. Another way to relax is through meditation. In 2022, researchers conducted a systematic review that examined 104 randomized controlled trials (studies in which participants are randomly divided into groups that compare different interventions). The analysis showed a higher ratio of positive outcomes for sleep (74 percent) and fatigue (68 percent) for those who practice meditation.[24] We explore the link between stress and mitochondria more in the next chapter.

Educate yourself and seek help if needed. If you don't feel well rested when you wake up, and if this happens regularly, you don't have to accept that you're a "bad sleeper"; it may indicate that you have a problem, and effective solutions are available. If problems persist, seek expert help, because there are sleep disorders that require professional intervention. For example, someone with obstructive sleep apnea won't benefit much from general sleep advice, because their airway is physically blocked. So people with consistent sleep problems should seek professional help for diagnosis and treatment. At the same time, have realistic expectations; meta-analysis of the effectiveness of sleep interventions show that the effectiveness varies across people and conditions, but overall, sleep can be improved.[25] NASA astronauts participate in an individually tailored education program that covers the importance of sleep and strategies for better sleep management.

• • •

Our mitochondria work so hard for us throughout the day that we must give them time to recover, and the best time to do it is at night. Think about it: Can you imagine running a factory around the clock without replacing broken parts and cleaning up the soot in the machines? Can you imagine sending the cleaning crew into a bustling office building during the day? The advantages of doing these activities after hours are obvious. By the same logic, it makes sense to give your mitochondria enough uninterrupted time at night to do their important cleanup work. Relax at night, and your mitochondria will thank you in the morning. And speaking of relaxation, it's important to find a way to do it throughout the day, and not just before you go to sleep, which takes us to the next chapter on avoiding long-term stress.

CHAPTER 10

Avoiding Prolonged Stress

On October 17, 1989, Daria was driving our old brown Volvo station wagon back from work at UC San Francisco to our home near Berkeley. Then, exactly four minutes after five p.m., when she was in the middle of the Bay Bridge, which connects San Francisco with the East Bay, she felt the car shaking. She held tight to the steering wheel, but something was pulling her to the left. Traffic on the bridge came to a halt. After a minute or so, she noticed some drivers who got out of their cars, talking excitedly and pointing ahead. A section of the upper deck of the bridge had collapsed. The radio reported that the Bay Area had been shaken by a major earthquake. "The kids!" was her first thought. She knew that our four kids were home with the babysitter. Are they okay? Did the house collapse on them? She noticed a driver in a nearby car who was talking on a cell phone (a rarity in 1989) and asked him if she could make a quick phone call. "I have four young kids at home," she explained, but the guy rolled up his automatic window and turned his head away. There was confusion everywhere. Some drivers left their cars and started running to get off the bridge, fearing that additional sections might collapse. Somebody said that the instructions are to stay in the cars. Helicopters searching for survivors from a car that plunged into the bay were further shaking the bridge. Around six thirty, darkness fell on the city. Daria and her carpoolers sat in the car quietly, waiting for further instructions from the police. They could see the fires in the Marina District of San Francisco. From time to time, they could hear the squeaking of the bridge.

Emanuel was two thousand miles east, at a trade show in Ann Arbor, Michigan, with a startup he had joined two years earlier, when someone told him that San Francisco had been hit by a massive earthquake. "The kids! Daria!" was his first thought. He ran to his hotel room, turned on the TV to images of burning buildings and live aerial images of a damaged Bay Bridge. Wild scenarios started running in his head as he tried to call home, neighbors, and friends, but all the lines were busy.

What happens to our body in stressful situations like this? Let's focus on two things that are part of the fight-or-flight response: One is adrenaline, which is secreted from the adrenal glands (that sit on top of your kidneys) to increase your heart rate. This ensures you have enough blood in your vital organs, mainly the skeletal muscles, that can be used to flee the source of danger or to fight your attacker. This hormone came in handy later that evening on the Bay Bridge, when police instructed the drivers who stayed in their cars to turn their vehicles around and get off the bridge. Unfortunately, there were some abandoned locked vehicles blocking the way, so sailors who came to help from a nearby naval station used their adrenaline-pumped muscles to lift those cars by hand and get them out of the way. Our Volvo was one of the last cars to get off the bridge.

The second important stress hormone is cortisol, which is produced in the mitochondria in the adrenal glands. Among other things, cortisol increases glucose levels in your blood and enhances your brain's use of glucose so that you can think.[1] Daria certainly needed this hormone to get off the bridge and then navigate in dark San Francisco, with fires and firetrucks all around. It was already close to midnight, but she had to be on full alert as she drove over the Golden Gate and then the Richmond Bridge and arrived home around midnight. The stress hormones helped her stay cool when she ran into an empty house, and then turned off the oven and the stove that had been left on. She drove up to the home of our good friend Adrienne Gordon, who had the kids, hugged them, and put them to bed. We both remembered that we had once agreed that Emanuel's mom would be our out-of-area contact, and this way, after a sleepless night, Emanuel got the news that Daria and the kids were safe. The following afternoon we were finally reunited as a family.

Your body's stress response is useful in emergencies, and it can help even in non-emergency situations: The stress that runners feel before a race can help them perform better, and the stress you feel before an important presentation may help you be focused and alert. Yet in the months that followed the 1989 earthquake, we discovered the ugly side of prolonged stress. Neither of us could shake off the experience. Yes, we handled the immediate crisis, but the stress did not go away. Before that earthquake everything seemed to be under control, and suddenly we realized that everything could be gone in a second. Daria would grow extremely anxious when driving under a double-decker bridge (forty-two people were killed during that earthquake when the upper level of a highway collapsed). Emanuel was constantly playing hypothetical scenarios in his mind. What if Daria had driven just a tiny bit faster and that 250-ton section that collapsed had crushed her car? In addition to the regular stress that comes with raising four kids and managing two careers, we both were constantly tired and easily irritated. We were super stressed and anxious. We are both self-reliant types who hate to ask for help, but when Daria started to cry one day during a coffee commercial on TV, she knew it was time to seek professional help. It took Emanuel longer to admit that he, too, could use it. In the end, we're glad we did, because this type of ongoing stress can be devastating to one's health.

There is mounting evidence linking chronic stress to poor health.[2] Why is ongoing stress so destructive? Because during a crisis, the body directs most of the ATP to the important stuff—the parts of your body taking care of the emergency—while limiting the ATP it spends on maintenance—on repairing the cell. This makes perfect sense. If your house is on fire, you'll focus on putting the fire out and helping others get out of the building. You won't be fixing the roof or checking for termites. These things could wait. But when you're always stressed, your body is always in crisis mode, and cell maintenance doesn't happen at the rate it should, because taking care of emergencies is always the top priority for our body. [3] It's as if the body says: "I'll take care of things when they're back to normal," but under chronic stress, they are never back to normal. In the months that followed the earthquake, when we were stressed and anxious, we were using energy that was diverted away from autophagy, mitophagy, and maintaining mitochondrial quality through balanced fusion and fission.

Stress is part of life, and it can be due to money issues, deadlines at work, relationships, and a million other things. We've all experienced stressful times, but problems start to emerge when it feels like the stress never stops. Constant exposure to cortisol and other stress hormones puts you at risk of (among other things) depression, anxiety, sleep problems, cardiovascular disease, and poorer immune function. So seeking help or other ways to reduce ongoing stress is important (and as we'll see in a moment, it can be done). Stress can also hinder recovery, as shown by research from Shanghai Jiao Tong University in China. The researchers there demonstrated that in mice with spinal cord injury, the stress hormone corticosterone (which is parallel to cortisol in humans) starts a chain reaction that negatively impacts the mitochondria and the energy transfer process. [4] This example illustrates the double whammy that is experienced by those of us who struggle with a disease or injury; as if having those isn't enough, the associated stress can delay recovery.

Stress from disease or injury often radiates to family members. Researchers from Ohio State University showed that wounds of family members who take care of Alzheimer's patients take an extra nine days to heal. [5] Mitochondria play an important role in wound healing, and while these researchers did not study the connection, it is possible that the delayed healing is linked to stress-related mitochondrial dysfunctions. Elissa Epel at UCSF has been studying stress for many years, and especially the long-term stress experienced by family members who take care of their relatives. These include, for example, relatives of people with mental illness, parents of autistic kids, or families of Alzheimer's patients. People in these situations often lack a sense of control over their lives, which is a major source of stress; our brain loves predictability, and lacking it causes these people to feel stressed, helpless, and overwhelmed.

After prolonged periods of caregiving, the health of these caregivers can be compromised. In work that Epel did with Nobel laureate Elizabeth Blackburn and others, they found that it also affects your aging and is linked to shorter telomeres.[6] Among women who cared for a chronically ill child, researchers found that the more years of caregiving, the shorter the mother's telomere length and the greater the oxidative stress. Notably, Epel and Blackburn reported that women who felt most stressed had shorter telomeres that on average were similar in length to women who were at least ten years older who felt low stress.[7]

How does all this relate to mitochondria? Around 2012, a young scientist named Martin Picard came to give a presentation at Epel's lab about his work on mitochondria. Picard, who had recently completed his PhD research on mitochondrial functions in skeletal muscles had become increasingly interested in the interplay of stress and mitochondria. Following his visit, Picard measured energy transformation capacity—what they called mitochondrial health—on some caregivers' blood samples that were stored in freezers in Epel's lab. When the analysis was done, the results were clear. The caregivers who have been under chronic stress for years had a lower mitochondrial capacity, which may explain why cells in such bodies prematurely age and are inflamed. And there was another finding in that study, one that perhaps should give us all some hope. In addition to the blood samples, Epel gave Picard very detailed questionnaires that participants had completed about their mood the night before blood was drawn from their arm. The analysis done by the researchers showed that elevated positive mood at night was associated with increased mitochondrial health in the morning.[8] In other words, if people reported that they felt closeness, love, or trust with other people around them, these feelings predicted better mitochondrial health the morning after. Picard says that more work needs to be done to study this correlation, but it gives us a clue regarding one way to protect our mitochondria from stress—putting ourselves in situations that we know have positively affected our mood in the past. It may also serve as a reminder that sleep and stress are closely related. This study didn't address it directly, but it's possible that the positive mood improved sleep quality and thus the mitochondria.

Not long after he gave the talk at Epel's lab, Picard started a postdoctoral fellowship in Doug Wallace's lab at Children's Hospital of Philadelphia. He remembers walking into Wallace's office one day and hesitantly suggesting an experiment in which they would use the laboratory's mice with abnormal mitochondrial components and then see how it influences their response to induced "mental" stress. Wallace, who had always been interested in the link between mitochondria and psychiatry, gave Picard the green light. There were five groups of mice in this experiment, one control group and four groups whose mitochondrial functions were altered in different ways. The researchers put each mouse in a tube with little holes to breathe but nowhere to go, which is a stressful situation. During the experiment, Picard and his colleagues monitored the level of corticosterone in the mice's blood (which, again, is parallel to cortisol in humans). After thirty minutes of stress, they released the mouse. The control group showed what was expected: The level of corticosterone climbed during the thirty minutes of stress, and this was followed by a period of recovery when the level kept going down. This was not true for the other groups. For example, in one group, mitochondria were modified so they could not transfer ATP to the rest of the cell. In that group, the level of corticosterone after the stress period was twice as high as the normal mice. The bottom line is this: This study showed that mitochondria played a role in how animals respond to stress. While this experiment was done on mice, Picard, now a professor at Columbia University, is currently studying this in humans, and he says that his initial results are consistent with the mouse study. Mitochondria may shape our responses to daily life stress.

Social Interaction: The Good, the Bad, and the Ugly

Some articles portray social interactions as the ultimate solution to stress, and it's true that a fun evening with friends can dramatically reduce your stress level. Reality, of course, is more complicated because other people can be a major source of stress. In this section, we'll look at this issue from different angles. Let's start with the good: In his book *Why Zebras Don't Get Ulcers*, Stanford neuroscientist Robert Sapolsky puts it this way: "The impact of social relationships on life expectancy appears to be at least as large as that of variables such as cigarette smoking, hypertension, obesity, and level of physical activity."[9] Numerous studies establish the link between social support and physical and psychological health.[10] Some studies have started to establish the link to mitochondria, too. For example, researchers at Duke University explored the link between social interaction and mitochondrial health in female rhesus macaques, and what they found surprised them. They studied nine social groups, each composed of five unrelated female monkeys. A member who is introduced later to the group has a lower status, and thus grooms more and is being groomed less. The researchers expected the dominant monkeys to have more mitochondria in their immune cells, but they found that it didn't matter if a monkey was groomed or if she groomed others. The number of mitochondria in their immune cells was positively related to *the time* females spent grooming either as the receiver or the giver.[11] In a nutshell, longer social interaction had a positive effect on mitochondria in this study regardless of the social status.

 -   In a 2024 study, researchers examined the relationships between several positive psychosocial factors, such as having a larger social network or having a greater sense of purpose in life, and the mitochondria in a certain part of the brains of four hundred people, aged sixty-five and older. This study was possible because the participants had agreed to donate their brains for research purposes, and when they were still alive, they reported on their different experiences. Caroline Trumpff from Columbia University and her colleagues analyzed the abundance of mitochondrial proteins in a region of the brain that is involved in emotion regulation and executive functions. They found a link between psychosocial experiences and the amounts of Complex I in the mitochondria; positive and negative experiences together were associated with 18 to 25 percent change in the amounts of Complex I in the brain cortex (the outer layer of the brain important for many high-level functions, including thinking and feeling). Single-cell studies showed that the effect was especially strong in glial cells—brain cells that support and protect neurons.[12] Of course, a sense of purpose in life can vary widely for different people, and it isn't exclusively linked to social relationships, but such relationships can play an important role in giving a person a sense of purpose.[13]

The opposite of social interaction is social isolation, and it has been associated with increased risk of developing heart diseases, type 2 diabetes, dementia, and other ailments.[14] Researchers at Rockefeller University showed that even fruit flies eat more and sleep less after about a week of social isolation.[15] How mitochondrial dysfunctions are linked to loneliness is yet to be explored, but some studies show that loneliness can lead to increased cortisol levels and overactivation of stress response, which ultimately cause mitochondrial dysfunctions. This in turn can create further mental and metabolic diseases.[16] The bottom line: Positive interactions with other humans are good for our health, and it's possible that mitochondria play a role in this phenomenon.

Yet seeing other people as the solution is only part of the story, because when it comes to stress, other people are often part of the problem: Conflicts at work, bullies at school, trolls on social media, difficult relationships, and drivers who cut you off in traffic can all contribute to your stress. How do these interactions affect our mitochondria? The following study starts to address this question. Imagine that you volunteer to participate in an experiment. You are given two minutes to prepare a short speech in which you'll have to defend yourself against a false accusation of shoplifting. After two minutes, a video camera is turned on and you are given exactly three minutes to deliver your speech while facing a man in a white lab coat. It is well established that public speaking causes stress, and if you've ever been falsely accused of anything, you know how stressful that is. Even though you know that this is part of an experiment, you most likely would display the typical symptoms of stress: your heart would beat faster, your muscles would tense up, maybe you'd feel some sweat on your forehead.[17]

This task was given to volunteers as part of an experiment at the University of Pittsburgh in 2011, and sure enough, participants displayed these typical stress symptoms. Thirty minutes after this brief stressful experience, blood was drawn from participants, and researchers detected a significant increase in pro-inflammatory state (which is obviously bad). A few years later, Martin Picard and his team at Columbia took a closer look at the blood samples, and they found something else that was floating in that blood: mitochondrial DNA.[18] As we explained earlier, mitochondrial DNA is found inside the mitochondria, but it is not supposed to flow freely in your blood, and when it does, it causes inflammation. We can deal with occasional inflammation, but if the stress is ongoing, if it happens chronically, it can lead to deleterious health effects like diabetes, heart diseases, and dementia.

Ongoing hostile interactions can be a major source of stress, and there are good reasons to believe that they can affect one's mitochondria. Imagine that you move to a new neighborhood, and on the first night you wake up to extremely loud music. You toss and turn, hoping that this is a one-time thing, but when the same thing happens the following night, you get out to look for the source of the noise. You ring the neighbor's doorbell, and when the door finally opens, you face a six-foot-ten-inch man who makes it clear that he's not going to change his ways because of the whimsical demands of a new neighbor. Neuropsychology researchers study these types of interactions in animals by placing a newcomer rat into the home cage of a more aggressive male "resident" for a limited time and assess the newcomer's reaction. It has been shown that repeated exposure to this type of social stress produces physiological and behavioral signs of stress. Specifically to mitochondria, this stress consistently impacts mitochondrial functions: It induces ROS, increases markers of oxidative stress, increases neuroinflammation, and reduces levels of antioxidant enzyme proteins.[19] Known among researchers as "social defeat," it also has been shown to reduce ATP production in certain brain regions and increase mtDNA mutations.[20] A study conducted in the U.K. found that bullying victimization in childhood predicts inflammation and obesity at midlife.[21] While dealing with bullies is beyond the scope of this book, recognizing that they affect your health is an important first step for finding a solution that can range from standing up to them to cutting ties and getting them out of your life.

Importantly, we each respond differently to stress. Some of us are inherently relaxed, while others are more worried and see things as threatening. Some scientists believe that those in the latter group with "trait anxiety" can be at a disadvantage, which leads to social subordination. In other words, they suspect that people who are usually more anxious can be predisposed to perform poorly in social competition. Carmen Sandi, professor at the Swiss Federal Institute of Technology in Lausanne, Switzerland, studied trait anxiety and social hierarchy and showed that rats with less anxiety were more likely to be dominant in a competitive situation. These less anxious rats had greater mitochondrial function in a part of the brain called the nucleus accumbens, which is important to motivated behavior.[22] She and her group categorized rats along a spectrum from low-anxious to high-anxious, and then introduced competition tasks to all the rats. Most high-anxiety rats took on lower social status. Their nucleus accumbens showed lower mitochondrial function. Essentially, highly anxious rats retracted earlier, whereas low-anxiety rats continued to fight longer for themselves.

The most fascinating part is that Sandi and her team were able to confirm their findings by either blocking or enhancing the mitochondria in the rats they studied. When low-anxiety rats received blocking agents, their competitiveness dropped and so did their social status. In contrast, when high-anxiety rats received enhancers, they performed significantly better and thus achieved higher social status. While rats are not humans, and social dynamics are very complex, Sandi and collaborators showed similar results in a study of 229 human subjects. Exposure to acute stress (using a procedure called the Trier social stress test) showed that low-anxiety individuals became overconfident, and high-anxiety individuals became underconfident. And similar to the rat study, cortisol responses to stress were higher in those with higher baseline anxiety.[23] These experiments are yet another reminder of the negative effects that chronic stress and anxiety can have, and why we should not accept them as a given. These studies may also highlight the importance of knowing yourself and developing strategies that fit you best. There might be a limit to how much anxiety trait can be changed, so people who are highly anxious may want to avoid too much stress, which can hurt their performance. Importantly, these individuals can sometimes perform better than others in a low-stress situation.

What Can Be Done to Reduce Stress?

Chronic stress is closely linked with poor health, and we've presented growing evidence that it is associated with mitochondrial dysfunctions. This should motivate us all to make sure we are not exposed to ongoing stressors or, if we are, to develop strategies to cope with them.

Reduce exposure to stressors. Certain sources of stress are unavoidable (there was nothing Daria could have done about that earthquake),[24] but there are stressors that you can control. If being on social media or watching the news makes you anxious, you can reduce your exposure to these information sources. You can still follow what's going on in the world, but you have the right to protect your body by being selective in what you watch and for how long. If you work for a toxic boss or are involved with an abusive person, we know this can be difficult, but your ongoing stress can drop substantially if you find another job or end the relationship. Sometimes, you can reduce stress by asking for help with your workload or giving up some responsibilities. Unfortunately, in many cases, things are more complicated, and the rest of this section deals with unavoidable stress. Still, it's worth stopping for a moment to ask this simple question: Is there a way to minimize my exposure to certain stressors to begin with?

Increase positive social interaction. There is clear evidence that positive social interactions are good for our health. In a way, we can all learn from our mitochondria how to be social. They don't miss an opportunity to connect with other mitochondria to exchange the latest news and meet new friends inside and outside the cell. They are part of a network, as we are, too. How can we become more social? Former U.S. Surgeon General Vivek Murthy, who sees loneliness as a pandemic that seriously impacts public health, offered a simple formula to getting started. He encourages people to take one simple action each day to express gratitude, to offer support, or to ask for help.[25] Here are a few more ideas: join a club, a writing group, a support group, a cooking class, a volunteer group. Exercise may also reduce stress, and doing this with other people can have additional benefits, so by joining a yoga or a salsa class, a hiking group, or a recreational adult soccer team, you reap a double benefit—the physical and the social. If you don't find anything you'd like to join, you may want to start a club: a book club, a nature club, a film club (we started one nine years ago and we love it). Community leaders can create opportunities for people to interact with one another by establishing things like street fairs, farmers' markets, or community gardens.

Invoke the relaxation response. The relaxation response (RR) can be seen as the opposite of the stress response. While the stress response is characterized by rapid heart rate and faster breathing, during the relaxation response one experiences slower heart and respiratory rates. RR occurs after practices such as meditation, repetitive prayer, yoga, tai chi, and breathing exercises. There is evidence that it reduces stress, and some initial evidence that shows the impact on mitochondria. In 2013, researchers at Harvard and the Massachusetts General Hospital recruited two groups of people: twenty-six subjects who had no prior RR-eliciting experience (the novices) and another group of twenty-six subjects who had significant prior experience of regular RR-eliciting practice. The novices then went through an eight-week training program that taught them some of those relaxation response practices. As the researchers put it in their report: "Our results for the first time indicate that RR elicitation, particularly after long-term practice, may evoke its downstream health benefits by improving mitochondrial energy production and utilization and thus promoting mitochondrial resiliency."[26] In other words, relaxing through practices such as meditation may improve the mitochondria.[27] One way to invoke the relaxation response is through deep breathing. With shallow breathing (the opposite of deep breathing) you might not supply enough oxygen to mitochondria throughout the body. In contrast, when you breathe deeply from the diaphragm, oxygen reaches all your cells, and you also get rid of more of the carbon dioxide that your mitochondria produce. While there is still limited evidence of their effect on mitochondria, the positive effects of meditation, yoga, deep breathing, and other relaxation methods on stress are well established.[28]

Spend time in nature. Spending time in nature has been shown to reduce cortisol level (again, this is the stress hormone produced in the mitochondria in the adrenal glands). There is a term in Japanese—*Shinrin-yoku*—which means forest bathing, and it refers to spending relaxing time in a forest. Meta-analysis of studies that compared the cortisol levels of groups that spent time in an urban setting versus groups that spent time in a forest showed that the cortisol levels were significantly lower among those who had spent time in a forest (i.e., practiced *Shinrin-yoku*).[29] Of course, it doesn't have to be a forest. Consider the following 2019 study conducted at the University of Michigan. Over a period of eight weeks, thirty-six participants were asked to spend time in nature at least three times a week. The researchers found that twenty minutes in nature significantly reduced the cortisol level in those participants, regardless of whether they were walking or simply sitting down.[30] A study in the U.K. found that spending 120 minutes (or more) in nature every week was associated with good health. (This survey included parks, beaches, the countryside, woodland, hills, and rivers.)[31] Unfortunately, some people cannot reach green spaces as easily as others. Community leaders and policy makers should take note because lower-income communities are less likely to have access to parks and other greenery. The link between stress reduction and overall health indicates that changing that should be a high priority.

Find opportunities to laugh. Think of the last time you laughed so hard that your coffee came out your nose or you sprayed your desk with soda. We're not sure what the coffee did to your nostrils (or the soda to your desk), but it was probably good for your stress level. Analysis from the University of Toronto of eight studies that examined the impact of laughter showed a significant reduction in cortisol levels by 32 percent.[32] In fact, it's possible that even just the anticipation of a good laugh may reduce cortisol levels.[33] In a small study reported in a 2008 scientific meeting by Lee Berk from Loma Linda University in California, the researchers told eight men that they were going to watch a humorous video that had been previously selected by those participants. Berk and his team were surprised by the fact that the cortisol level in these men decreased by 39 percent even before they watched the video. Less surprising (but very encouraging) was the fact that their cortisol levels decreased 67 percent thirty minutes after they watched the videos. This latter finding is consistent with other studies that show the stress-reducing effect of laughter. So look for opportunities to laugh (but you may want to put down your drink first).

Accept what you cannot change. As we mentioned, our brain loves predictability, and lacking control is a major source of stress. However, some things in life are not under our control, and being unable to accept a new reality can be harmful as has been shown in studies with monkeys. Monkeys who are dominant in a group are healthier than their subordinates, but life happens for monkeys, too: conflicts with other groups, challenges within the group, extreme weather, or other external factors may change the social order and cause a previously dominant monkey to find himself as a newcomer in a different group. Researchers found that such a monkey is more stressed and is likely to develop more extensive atherosclerosis (hardening of the arteries) than subordinates[34] (and mitochondrial dysfunction has been associated with atherosclerosis).[35] It's hard for that monkey who used to be the boss to accept the new unstable social environment, which stresses him and affects his health. For us humans, there are many examples of things that are beyond our control: a new boss at work, the death of a loved one, sickness, natural disasters, and economic fluctuations are just a few examples. Accepting things as they are can be hard, but it can greatly help us reduce stress and reach peace of mind.[36]

Work on multiple fronts. Again, as we pointed out earlier, links have been established between psychological stress and things like sleep and the microbiome. And physical activity has been shown to reduce stress, too. By taking care of these parts of your life, you may increase your resilience and thus your ability to deal with external stressors and bounce back after a difficult experience. The flip side of this is that when you let one of these elements slip, you may create a vicious cycle. Consider sleep, for example: When you are stressed, you don't sleep well, and when you don't sleep well, your mitochondria aren't happy, and they may be contributing to your feeling stressed. But vicious cycles are stoppable; you don't have to be *perfect* on all fronts to be healthy, and sometimes a small improvement in one area can break the cycle.

Seek help. Theoretically, we can solve our mental problems on our own, except for one obstacle: We're often locked into a certain way of thinking, which had gotten us where we are in the first place. It's impossible (or, at the very least, extremely hard) to escape that way of thinking, but a professional therapist can offer a different perspective and may have some tools to help. Whether it's psychotherapy or medication (or both), professionals can help treat anxiety and stress. Covering the variety of tools available is obviously beyond the scope of this book, but just one example is cognitive behavioral therapy, which can be effective in reframing how we think of something and has been shown to reduce anxiety and stress.[37]

●　●　●

When you're stressed, you literally waste energy,[38] thus exhausting your mitochondria, so think of the last time you were truly relaxed. When was it and what were you doing? And what are other things that help you relax? In addition to the stress reduction guidelines listed above, it's worth identifying the things that calm you down. This could save you energy and allow your mitochondria to power health-promoting processes rather than expensive stress responses.[39] For example, we know that we are hardly ever stressed after a long walk on the beach, after dinner with friends, when listening to music, during or after a visit with our kids and grandkids. Like everyone else, our lives are not free of stress, and we would be lying if we said that we can easily brush off its effects, but we are very aware of what helps us relax, and we consciously repeat those things as often as possible. Of course, there are things that reduce stress for only one of us, so we do them separately. For example, Daria finds doodling or shopping for shoes extremely relaxing. Emanuel is never stressed after his daily nap, or after twenty minutes on his stationary bicycle. Likely, your list will be different, but once you identify it, follow this simple formula: Enjoy. Relax. Repeat.

CHAPTER 11

Avoid These to Help Your Mitochondria

Everyone knows that cyanide can be fatal, but have you ever wondered how it can kill someone? It poisons the mitochondria by binding to Complex IV in the electron transport chain.[1] Once the mitochondria come to a halt, essential organs like the brain and heart stop functioning almost instantly. Yet as dangerous as it is, there's one piece of good news about cyanide: Everyone knows that it can be fatal. What about poisons that we don't know about and therefore we fail to avoid?

As you will see in this chapter, things that might harm our mitochondria can hide in surprising places: a picturesque farmland, a barbecue pit, the supermarket, or on a boat. Today, it is almost impossible to eliminate all exposure to toxins. So our goal is to give you information that will help you make educated choices based on your preferences as you try to reduce your exposure. The main point is that we want to minimize the extra tasks we give our mitochondria. It's bad enough that they must constantly clean ROS and other toxic metabolites that accumulate as part of the mitochondria's normal operation, so we certainly shouldn't overwhelm them with additional burden. This is especially important as we get older and our ability to undo damage to our mitochondria dwindles, our reserve is lower, our repair machinery is weakened, and irreversible damage to the mitochondria accumulates. In this chapter we'll discuss research on the impact of things like smoking, vaping, alcohol, air pollution, radiation, fried or barbecued food, exhaust fumes, and pesticides. By limiting such potentially harmful agents, we reduce damage to the overworked mitochondria, and thus, we improve the health of the cell.

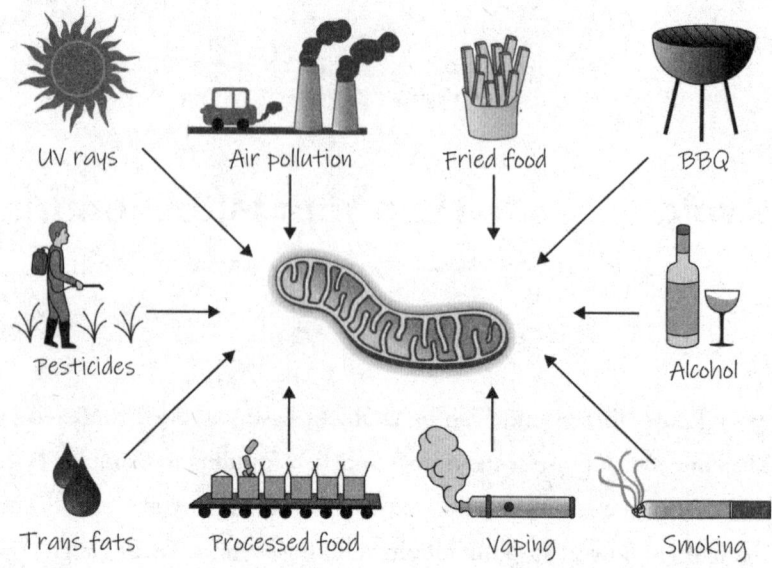

UV rays   Air pollution   Fried food   BBQ

Pesticides   Alcohol

Trans fats   Processed food   Vaping   Smoking

Potential harmful agents to the mitochondria: alcohol, smoking, vaping, processed food, trans fats, pesticides, UV rays, air pollution, and fried and barbecued food.

Damage from Alcohol, Smoking, and Vaping

It is well known that excess consumption of alcohol can be bad for your health, but most people are not aware of the mitochondria's involvement in this. As we briefly mentioned, alcohol metabolism (that occurs mainly in liver cells) generates a toxic molecule called acetaldehyde that wreaks havoc in the mitochondria and in the cell. Acetaldehyde-induced damage happens even when you drink one glass of wine or a beer, but this damage is often temporary. Your mitochondria usually help metabolize acetaldehyde to non-toxic metabolites, remove damaged cell parts, and thus restore the cell to normal function. We said "usually" because if acetaldehyde is not metabolized fast enough, it directly binds to proteins, lipids, and DNA, and it is harder for the cell to get rid of the resulting product. In the long run, habitual alcohol consumption results in accumulated damage to many cellular and mitochondrial components. To understand why, think of a cleaning crew that comes to clean up an office after hours. Normally, they finish the job in a couple of hours. One day, the office workers have a party before they go home at five p.m., and they leave a mess behind. That evening, it takes the cleaning crew four hours to clean up, leaving them especially tired. Not good, but not the end of the world if it happens once. However, when those parties occur again and again, the cleaning crew is exhausted. This is reflected in their performance, and the office isn't as clean as it used to be. There's more: Sometimes, the party gets so wild that rowdy office workers attack the cleaning crew, further affecting their performance. A key member of the cleaning crew is a mitochondrial enzyme called aldehyde dehydrogenase 2, which is constantly removing aldehydes. However, this enzyme, too, can get exhausted. Sometimes it is even irreversibly inactivated by aldehydes, and replacing it is a task that requires more ATP that is diverted away from other cell functions.[2]

What about everything that we heard over the years about the health benefits of moderate drinking? The pro-alcohol view has shifted dramatically in recent years, and many scientists believe that no amount of alcohol is good for you. We realize that hearing conflicting recommendations can be frustrating, but advice should be based on evidence that sometimes changes. For a few decades, many scientists believed that small amounts of alcoholic drinks were good for our hearts.[3] For example, scientists who supported the so-called French Paradox theory pointed out that despite a high intake of saturated fat, the French had lower rates of heart disease than people in other countries and attributed it to their consumption of red wine. (Seventeen years ago, Daria, too, showed that applying a small amount of alcohol to rat hearts, just before a simulated heart attack, reduces the damage to heart muscle cells.[4]) However, in recent years, there has been a shift away from the pro-alcohol view that has resulted in new guidelines. For example, in January 2023, the World Health Organization stated that "no level of alcohol consumption is safe for our health," stressing that the risks of drinking alcohol start from the first drop.[5]

The American Cancer Society (ACS) states that it is best not to drink alcohol, explaining that "to reduce the risk of developing several types of cancer, there is no safe level of consumption." For people who still choose to drink alcohol, ACS and other health agencies recommends limiting consumption to no more than one drink per day for women and no more than two drinks per day for men. (Women tend to be smaller and to break down alcohol more slowly.)[6] In January 2025, the U.S. Surgeon General, Vivek Murthy, argued that alcoholic beverages should carry a warning label stating that alcohol is a leading cause of cancer.[7]

The combination of smoking and alcohol seems to be particularly dangerous. Both increase aldehydes, including acetaldehyde, and both are known risk factors for upper digestive tract cancers (cancers of the mouth, esophagus, stomach, and small intestine). Researchers at Helsinki University Central Hospital demonstrated the increased risk by measuring the amount of acetaldehyde in the saliva of two groups of volunteers. One group consisted of people who both drink alcoholic beverages and smoke regularly. The second group consisted of people who just drink regularly. The first group (those who smoked *and* drank) had a huge rise in acetaldehyde in the mouth (more than tenfold) after each cigarette.[8]

Smoking that is not coupled with alcohol consumption is dangerous, too, of course.[9] Both our fathers hardly ever drank alcohol, but were heavy smokers, and both died of heart disease at a young age—Daria's dad when he was sixty-four, and Emanuel's dad when he was just thirty-six years old. Why is smoking dangerous? Smoke from tobacco contains hydrogen cyanide as well as carbon monoxide, acetaldehyde, acrolein, and formaldehyde, all of which negatively affect the mitochondria.[10] And of course, there's nicotine, a toxic chemical for the heart. By damaging mitochondria, smoking generates reactive oxygen species (ROS) such as peroxides, superoxide, hydroxyl radical, singlet oxygen, and alpha oxygen, all of which generate additional aldehydes when interacting with other components of the cells. You can think of these aldehydes as a bunch of criminals moving to your neighborhood. To fight them, the cell and its mitochondria have police-like enzymes that break down those toxic reagents, but these helpful enzymes can go only so far in getting rid of the bad players. This is particularly a problem if the damage occurs in the mitochondria. Recall that a heart cell muscle is tightly packed with mitochondria because pumping blood requires a lot of energy. So when some mitochondria in the heart muscles are damaged, there is a limited supply of energy, and the muscle doesn't operate at full capacity.[11] While we don't have all the details of our fathers' conditions, it's possible that at least some of their heart malfunctions were caused by aldehydes generated by the smoke they inhaled. Years after our fathers died, Daria's lab set up to help these "police-like" enzymes that get rid of aldehydes. The drug that they identified (called Alda, after Daria's mother) greatly increased the removal of the aldehydes that accumulate during a heart attack in a rat model, thus reducing the damage to the animal's hearts.[12] Whether this drug will be beneficial for humans remains to be determined. (Disclosure: Daria holds patents related to Alda.)

The heart is not the only organ affected by smoking. Every tissue that those aldehydes touch will be damaged as they bind to lipids, proteins, and DNA. Smokers increase the concentration of aldehydes in their mouth and nose, which are rich in blood vessels, so it is very easy for those aldehydes to get through the bloodstream into other parts of the body. Through the nose, the brain gets an extra speedy delivery of aldehydes, which can trigger neuroinflammation and neuronal cell death. Smokers' skin is oftentimes affected, too; they have more wrinkles, which reflects the fact that they had a high concentration of aldehydes that damaged skin cells. The high concentration of aldehydes in their saliva put smokers at a higher risk of cancer in their mouth, and since they swallow the saliva, smoking is also a risk factor for cancers of the esophagus, stomach, and small intestine.

Cigars, waterpipes (hookah and narghile), and electronic cigarettes (vaping) can affect mitochondria, too.[13] In 2021, researchers at the University of Southern California led by Ahmad Besaratinia showed that, like cigarette smoking, vaping by itself is linked to dysregulation of mitochondrial genes and increased inflammation.[14] Researchers at Rutgers University showed that electronic cigarettes induce mtDNA damage and may lead to atherosclerosis.[15]

Irfan Rahman at the University of Rochester explains that mitochondria play an important role in healthy lung function, and the chemicals users inhale when vaping can create permanent damage. "Eventually, such mitochondrial disruption produces inflammatory responses that can increase lung aging, so that a forty-five-year-old who vapes might actually have the lung function of a sixty-year-old," he said in a recent interview.[16]

Damage from Food or the Way It Is Prepared

Consider a guy named Joe who's totally into barbecuing. If you asked Joe on Sunday night how his day went, he would say it was just perfect. Standing by the grill, flipping burgers, steaks, and hot dogs for his extended family is one of his favorite things. He loves the sizzling sound of meat touching metal, the smell, the beer, the crunchy French fries his wife makes in the kitchen, and of course the compliments from everyone. On Sundays, Joe is in heaven. Yet if you asked Joe's mitochondria about their afternoon, you may get a different answer, because there were several things that afternoon that could have hurt them.

First, like cigarette smoke, the smoke that comes from grilling is bad for mitochondria, and anyone who ever barbecued knows the feeling of finding themselves in a cloud of smoke when the wind suddenly changes. Inhaling smoke can be dangerous because it contains the odorless gas carbon monoxide, which can slam on the brakes of mitochondria just like cyanide does (more on this later). Smoke from BBQ also contains harmful chemicals called polycyclic aromatic hydrocarbons (PAH), which are a class of chemicals in coal, crude oil, and gasoline.[17] Like other toxic agents, PAHs generate ROS and therefore damage the cell. PAHs and another compound called heterocyclic amines increase the risk of cancer and other chronic diseases. Importantly, you can be affected by PAHs not only by breathing them in; eating grilled or charred meat with PAH particles that have settled from the air is another way that you are exposed. Joe isn't the only one who can be affected, so are his family and friends who have enjoyed his delicious food.

Are we saying that Joe and his family should never barbecue again? This is up to them to decide, of course, but knowing the risks is the first step, because then they can consider ways to reduce it. For example, they can barbecue less frequently or trim all visible fat before grilling (fat tends to drip on the flame and produce more smoke with PAHs). Joe can wrap the meat in foil, which will protect it from smoke, he can move the meat away from the hottest part of the grill after the initial searing, and he can remove charred parts of the meat before serving it. Joe and his family can also substitute meat with fish and vegetables that don't produce as many PAHs.[18]

We don't want to spoil anyone's fun, so please forgive us, and keep in mind that we're trying to represent the voice of trillions of tiny life machines—mitochondria—that have been silent for way too long. Case in point—processed food. We personally love deli meats, potato chips, cookies, and other things that fall under this category, but over the years we have cut down on many of these because we know that they are bad for our life machines. In addition to the problems of ultra-processed food that we discussed in previous chapters (high salt and sugar content and low fiber), processed food often contains a class of compounds called advanced glycation end-products (AGEs), which make food taste delicious but are bad for our mitochondria. Glycation is a chemical reaction that happens during food preparation, like baking, cooking, and frying when sugars react with proteins or fats.[19] Like aldehydes, AGEs can trigger cancer when interacting with DNA, trigger inflammation when interacting with proteins outside the cell, and inhibit cellular processes when interacting with proteins inside the cell. In addition, they can make membranes leaky, and all these interactions contribute to mitochondrial dysfunctions and other problems associated with chronic diseases.

Is there a way to tell if the food you eat contains AGEs? Unfortunately, this is not something that appears on the food label, but here is a simple guideline. When you eat raw foods (like fresh vegetables and fruit) or steamed or boiled food, you reduce your exposure to AGEs. In contrast, when you eat roasted, grilled, or processed food, you increase your exposure. Anytime that you see that golden-brown color in baked goods or in caramelized sugars, this is a sign of a high level of

AGEs. The same is true for when you taste that roasted flavor in grilled or fried meat. As always, quantity matters, and in our personal life we haven't eliminated these delicious things, but we have significantly reduced them.

Trans fats are a category of ingredients that can increase your "bad cholesterol" (LDL) and lower your "good cholesterol" (HDL), and in recent years evidence emerged that they negatively affect the mitochondria. For example, in 2020, Japanese researchers found a link between some trans fats and several disorders, including cardiovascular and neurodegenerative diseases. They found that trans fats can enhance mitochondrial signaling that leads to programmed cell death.[20] Artificial trans fats are bad for you. Period. They are created during a food-manufacturing process that was discovered in the late nineteenth century but since has proven to have serious deleterious effects. It is tricky, though, to figure out what foods contain trans fats. While they have been banned in certain parts of the world, they are not banned in others. For example, if you live in a developing country, or travel to one, keep in mind that the use of trans fats is still very common there. In the United States, where they have been banned, small amounts are still allowed in food. Annoyingly, if the food has less than 0.5 grams of trans fats per serving, the food label can read 0 grams trans fats, and all these trans fats can accumulate in your body.[21] What to do? Look at the label. If it lists "partially hydrogenated oil," there's trans fat in the food. Foods that are likely to contain trans fats include commercial baked goods, microwave popcorn, frozen pizza, fried foods like French fries, doughnuts, and fried chicken.[22]

Speaking of fried foods, they can hurt mitochondria in other ways. Oil at high temperatures can form aldehydes that have serious deleterious effects.[23] Those aldehydes accumulate (especially if you reuse the oil), they may enter your lungs, and they can penetrate the food you fry. One of those aldehydes is 4-HNE, which can bind to proteins, thus changing their function and stability. 4-HNE can also increase mitochondrial ROS, reduce mitochondrial functions, and activate programmed cell death.[24] Another one is acrolein. The name derived from a combination of two words—acrid and oleum— because of its disagreeable odor and oil-like consistency.[25] When you smell very hot or smoky oil, it means that a tiny number of molecules of this toxic compound reached your nose, signaling you to avoid the source of the offensive odor.[26] Some acrolein also goes into the fried food you'll eat, and it eventually can reach the cell.

Not all researchers are equally concerned about the danger of deep frying. Consider two meals we had with two scholars. One was with Paolo Bernardi, a veteran mitochondria researcher who insisted that we should try one of his favorite dishes. Paolo drove us for a couple of hours to a restaurant in northern Italy, and when the waiter arrived, we each had a deep-fried wheel of cheese the size of a plate in front of us. It tasted and smelled wonderful. Daria ate only a few bites, but Emanuel couldn't resist the deliciousness. The other meal was in California with Tetsumori Yamashima, a Japanese physician and neuroscientist who argues that you should throw away all the omega-6 cooking oil you have in the house and never consume anything made with it.[27] We don't remember what we had for dinner with Tetsumori and his lovely wife and daughter, but it wasn't deep-fried cheese or anything else deep fried. While we are not as extreme as Tetsumori, and we can't resist French fries from time to time, we generally follow his and others' recommendation to use extra virgin olive oil for low temperature frying (sautéing).[28]

Damage from Pesticides, Air Pollution, and Radiation

Exposure to pesticides can be harmful to mitochondria, too. For example, there is strong evidence linking exposure to pesticides and Parkinson's disease. You may recall the impressive detective work of the neurologist William Langston after he was called to solve the mystery of young frozen patients with Parkinson's-like symptoms. As we described in chapter 4, it turned out that all these patients were exposed to heroin tainted with a compound called MPTP, which started a domino effect involving the mitochondria, and damaged neurons in an area of the brain that helps coordinate movement. Not long after that finding, another fascinating discovery was made: The chemical structure of MPP+ (the toxic metabolite of MPTP, which can cause Parkinson's disease) is strikingly similar to that of a widely used herbicide called paraquat.[29] This led scientists to investigate the link between Parkinson's disease and chemicals used in agriculture, and they made some alarming discoveries.

For example, a 2011 analysis conducted at the Parkinson's Institute (which was founded by William Langston), found that paraquat exposure increased the risk of Parkinson's disease by 2.5-fold among farmers and their families.[30] The risk, however, is not limited to agricultural workers because paraquat and other chemicals can drift in the air or contaminate groundwater. Analysis by researchers at UC Berkeley found that exposure to paraquat (and another chemical—maneb) within five hundred meters of one's home increases the risk of Parkinson's disease by 75 percent.[31] What about the fruits and vegetables sprayed with paraquat? It is challenging to establish a link with Parkinson's and the consumption of fruit and vegetables that are sprayed with pesticides. Thoroughly washing produce is always recommended.[32] Another chemical, rotenone, which was used as an insecticide, also increased the risk of Parkinson's disease significantly. These chemicals harm the mitochondria by inhibiting Complex I and by generating ROS.[33]

Paraquat has been banned in many countries, including Sweden, Germany, and China. Yet despite ongoing efforts of scientists and activists to ban it in the United States, it is still being widely used on cornfields, soybeans, wheat, cotton, and grapes.[34] Another substance that has been shown to negatively affect mitochondria—DDT—has been banned from use in the United States since 1972, but there are parts of the world where it is still being used, and in the United States, DDT (and its metabolite DDE) still show up in people's blood decades later.

Air pollution is another threat for mitochondria, and we will focus here on two types of pollutants. The first one is particulate matter that can originate from car emissions, industrial processes, and other sources. It is a mixture of chemicals, both solids and tiny liquid droplets, that float in the air. These particles are so tiny that they are suspended in the air and don't fall to the ground, and when we walk around, we can inhale them. Even repeated low-level exposure to air pollution can be harmful to the mitochondria. An Australian study showed that after three weeks, mice exposed to low levels of traffic-related particulate matter pollution had an inflammatory response that exerts effects even at low concentrations. "In our model we found strong, and statistically significant evidence of lung inflammation and dysregulated mitochondrial activity,"[35] the researchers wrote.

The second example of air pollution is carbon monoxide (CO), which can be emitted from cars, industrial processes, wildfires, and other sources. As with other pollutants, carbon monoxide, too, can raise its ugly head in unexpected places, even during a quiet day on a calm lake. Consider, for example, what happened to one Oklahoma family in 2020. It started as a happy boating day on Lake Eufaula in Oklahoma. Nine-year-old Andy Free had fun with his siblings and parents on their family boat—they did some swimming and tubing, and a few hours later headed back to the dock. It was then that Andy fell into the water. He was pulled out; CPR was performed but to no avail. The assumption was that Andy drowned, but the coroner's report indicated a high level of carbon monoxide in Andy's blood.[36] It turned out that Andy spent time at the rear of the boat, and he inhaled enough carbon monoxide from the boat's engine to cause serious damage to his body. Carbon monoxide affects mitochondria in a somewhat similar way that cyanide does—it essentially shuts down oxidative phosphorylation—the process by which ATP is produced.[37]

Most cases of carbon monoxide poisoning occur indoors. According to the Center for Disease Control, each year more than one hundred thousand Americans visit the emergency room after exposure to this odorless gas and more than fourteen thousand are hospitalized. More than four hundred die from unintentional carbon monoxide poisoning, and this does not include occurrences linked to fires.[38] Cases are largely preventable since the most common sources of carbon monoxide poisoning are well known: unvented space heaters, fires, clogged chimneys, exhaust fumes, malfunctioning cooking appliances, water heaters, furnaces, and clothes dryers. As Andy's case demonstrates, CO poisoning can happen outdoors, too, from generators that run on gasoline, motorboat exhausts, car exhausts, and campfires. Though some of these pollutants may be unavoidable, we should do our absolute best to minimize our exposure both inside and outside.[39]

Radiation is also harmful to mitochondria, and of course, the sun is the biggest source of UV radiation.[40] We don't want to vilify the sun, which enables life on Earth. Without the sun, mitochondria wouldn't have the two things they need to produce ATP: energy stored in food and oxygen—both made possible by photosynthesis in plants. That being said, it is well established that ongoing exposure to the sun can cause nuclear and mitochondrial DNA damage and oxidative stress in skin cells, which can lead to aging of skin cells and skin cancer.[41]

Your dermatologist is not kidding when they talk about protecting yourself from the sun because exposure to UV light is a major risk factor for skin cancer. Whether it's sunscreen, a hat, or a parasol (Daria's personal favorite), these things really work. And they protect your vulnerable mtDNA as well: One study reported a tenfold higher rate of deletion in a certain part of the mtDNA in cells that were exposed to UV radiation when compared to cells that were protected from radiation.[42] For a study that demonstrates the effect of UV on the aging of the skin, researchers in Germany recruited fifty-two volunteers whose "buttock skin had not been exposed to natural or artificial UV radiation for a minimum of 1 yr."[43] The researchers then repetitively exposed previously unexposed skin to UV radiation, and found that the frequency of mtDNA deletions in the exposed areas were approximately 40 percent higher than in the non-irradiated skin. Bottom line: Exposure to UV damages mitochondria.

What Can Be Done to Reduce the Risk?

Since you live on planet Earth in the twenty-first century, you can't completely avoid all toxins, but there are things you can do to reduce the burden on your life machines (which have their own pollutants to take care of).

Minimize damages. Some harmful agents are so dangerous that you should avoid them at all costs. For example, exposure to carbon monoxide can be fatal, so you want to do everything in your power to make sure you and your loved ones are protected by doing things such as installing detectors and verifying the proper operation of appliances. There are also harmful agents that come from things that are hard to give up. The key point to remembering is that numbers matter. In the end, it's about how many particles of a certain toxin will interact with healthy parts of your cells. If you decide that you don't want to give up a certain behavior, you can reduce their negative impact in three ways:

1. **Reduce frequency.** There is a huge difference between barbecuing twice a week or doing it once a month. The same is true for things like drinking wine and eating fried food or processed food. Cutting back will reduce the risk.

2. **Consider alternatives.** Barbecuing vegetables is likely to still generate toxins such as PAHs, but fewer of them. However, make sure the alternative is safe. For example, switching from smoking to vaping is not advised because, as mentioned, vaping, too, can hurt your mitochondria.[44]

3. **Use protection.** When applying pesticides and other chemicals, using protective gear will make a difference in your level of exposure. Protecting yourself from UV radiation from the sun is critical, of course. We always monitor the air quality, and when it's low (as happens frequently during wildfire season in California), we close windows, use air filters, and wear appropriate masks when we go outside.

Of course, giving up harmful behaviors is better than just reducing them. And you can further help reduce the exposure to pollutants by influencing policy, and by staying informed.

Seek help to change habits. If you decide to quit smoking or drinking, you don't have to do it alone. In fact, you're likely to increase your chance of success substantially if you use the help of professionals or certain support groups. Studies of smoking cessation show higher success rates when people used behavioral counseling as opposed to attempting to quit without help. Sometimes getting support from others who try to achieve the same goal is even more effective. For example, a 2020 analysis of studies on the effectiveness of Alcoholics Anonymous concluded that it helped more people achieve sobriety than therapy did.[45] Maybe the best piece of advice about bad habits comes from a guy on your $100 bill, Benjamin Franklin: "It is easier to prevent bad habits than to break them." Many researchers say this about alcohol these days: "If you don't drink, don't start."[46]

Influence policy. By promoting practices such as walking instead of driving (which is good for your mitochondria anyway), you can improve air quality for everyone.[47] By encouraging smokers to quit, you can affect how much cigarette smoke is in the air. And policy changes can make a difference: Analysis shows that raising the price of cigarettes by 10 percent increased smoking cessation rates by 3 to 5 percent, and that comprehensive laws to promote clean air indoors increased the rate of quitting by 12 to 38 percent.[48]

Stay ahead of the curve. Scientific studies usually precede public policy. For example, by the time the surgeon general issued the first warning against smoking, there were already seven thousand articles about the subject.[49] So keeping track of science is a good idea. Regulation proposals often face fierce opposition from industry, which slows down or prevents their adoption. Watching what happens in other developed countries can help, too. For example, Ireland, which was the first country to ban smoking in indoor workplaces, passed a law that requires a cancer warning label on alcoholic drinks starting in 2026.[50]

•   •   •

We all work hard to keep our homes free from potential invaders—burglars, water leaks, and pests are just three examples. It's easy to lose sight of the fact that, like our homes, mitochondria are physical places, too. Granted, they are tiny places, but they are places, nevertheless. And all these harmful agents with odd names—aldehydes, AGEs, PAHs—are real invaders that can enter our mitochondria or harm them in different ways. The name of the game is to keep those potential invaders away. By avoiding alcohol, smoking, vaping, processed food, trans fats, and fried and barbecued food, we limit their chance to get in. By protecting ourselves from pesticides, UV rays, and air pollution, we block them from getting close. All this with one goal in mind: to protect those magnificent life machines.

Hope Is in the Air

The man in front of us was lying on his back, his left arm stretched toward his shoulder, as if he were still trying to stop the blood gushing out of his body. He was a short man, around the age of forty-five, with noticeable lined tattoos on his ankle. We thought about his last moments: the blood with millions of platelets ready to start the coagulation process that would stop the bleeding, the mitochondria that came in those platelets to provide some of the building blocks for new blood vessels to replace the severed ones. Alas, the blood vessel that was hit—the subclavian artery—was too wide, and the blood never had a chance to clot. He probably turned pale as his body tried to direct the little blood that had been left to the most essential organs, but then these organs were starved of oxygen as well. With no oxygen to bind to electrons in his mitochondria, his life machines came to a halt: in his kidneys, his lungs, his heart, his brain.

We arrived in Bolzano, Italy, on September 20, 2024, exactly thirty-three years after the criminal investigation in this man's case was opened. He was found in the snow not far from the Italian Austrian borders by a couple from Germany, and some assumed he was a mountaineer who had disappeared some fifty years earlier. Five days after the body was found, an archaeologist was brought in, and he determined very quickly that the murdered person was ancient. Further lab tests estimated the body to be fifty-three hundred years old, and X-rays indicated he was hit by an arrowhead. The body was naturally mummified and gave scientists a rare opportunity to conduct studies on such an ancient corpse. Although he was officially named "the Iceman," the nickname that stuck was Ötzi, in reference to the nearby Ötz valley where he was found.

We hadn't planned to visit Ötzi. We were in northern Italy to participate in a three-day symposium in Padova to honor the legacy of our friend Paolo Bernardi, a veteran mitochondria researcher.[1] In 2016, we spent three months in this city, as Daria visited Bernardi's lab to learn the inner workings of oxidative phosphorylation, and we became good friends with Paolo and his wife, Rita. On the occasion of his retirement, a conference was organized with some leading mitochondria researchers from around the world, and we thought that this would be a great opportunity for us to catch up with our friends and hear about cutting-edge research in the field.

Since we had been working on this book intensely with hardly any time off, we decided to precede the conference with a week of hiking, "forest bathing," and relaxing up in the Dolomite mountains. During the weeks leading up to the trip, we came across the South Tyrol Museum of Archaeology, which is Ötzi's current address, and were reminded of his sensational discovery in 1991. In the hectic days that followed, we kind of forgot about Ötzi and his museum. We ate, slept, and breathed mitochondria, and what does a guy who's been dead for millennia have to do with the life machines? But on the way to our hotel up in the mountains, a travel brochure reminded us of him, and we started to wonder: What is known about Ötzi's mitochondria? Is it possible that one of us (or both) is a descendant of Ötzi, especially Daria, with her Italian roots? How similar are we to this man, and what can we learn from him today?

The small village where we stayed outside Bolzano was surrounded by forests of spruce trees and hiking trails. From the moment we arrived, we felt as if we had stepped into a documentary titled "Life as It Should Be Lived." We left our luggage at the hotel and went for a two-hour hike. The air was crispy clean, and we could almost feel our cortisol level drop as we strolled by green meadows, greeted by some local cows and horses, and faced the stunning view of fifty shades of green. The menu at the hotel restaurant included items like "fermented salad," and tips about things like eating whole carbs and the importance of antioxidants to protect against free radicals. There was no TV in our room, so we went to sleep even earlier than usual. We were in mitochondria heaven. Our possible connection with Ötzi was still on our mind, though, so the next day we searched online and found an issue of the scientific journal *Current Biology* from 2008 that featured our mummified friend on its cover with the headline: "Ötzi's Mitochondrial Genome." The article reported that Italian and British scientists analyzed cells taken from his intestines and that his complete mitochondrial genome analysis revealed that his haplotype cannot be found in people living today.[2] (Recall that a haplotype is an inherited mtDNA sequence unique to a specific ancestry group.) This meant that neither of us is a descendant of Ötzi's on our mothers' sides.[3] But during our search, we came across more interesting facts about Ötzi and became fascinated with the guy. A couple of days later, we took the short bus ride down to Bolzano and visited the museum.

Seeing some of Ötzi's belongings made him more relatable. His backpack looks like the one used by modern hikers, and his fur cap is not unlike those still used in cold countries. (Incidentally, thanks to mitochondrial DNA analysis, we know that his cap was made of bear skin, and that he also used goat, sheep, and deer as sources for his clothing and shoes.[4]) Five thousand years may look like a long time, but compared to the bacteria that was engulfed by some ancient cell 1.5 billion years ago to eventually evolve into mitochondria, Ötzi is not ancient at all. And the more we learned about him, the more we realized that Ötzi was so much like all of us. We think, for example, of pollution as a modern phenomenon, but researchers detected lots of soot in Ötzi's lungs, which were blackened, likely due to spending long hours near open fire. Certainly not good for his mitochondria. Most importantly, we learned that, like all of us, Ötzi wanted to get better. There are indications that he suffered from pain in his back and legs, and some scientists believe that his tattoos (sixty-one of them) served mainly as an early form of acupuncture to deal with his pain. We saw the pieces of birch fungus threaded on a string that he carried with him and are believed to have been used for therapeutic purposes.

We've come a long way since Ötzi's days. Medicine science has made tremendous progress, especially in the past one hundred years: the invention of antibiotics in the 1940s, the advances in genetic engineering and the birth of biotech in the 1970s, obtaining the complete human genome in the 2000s, cell therapy in the 2010s, and more. Will the breakthroughs in our understanding of mitochondria's pivotal role in our health lead to such a milestone?

Mitochondria and the Future of Medicine— A Hopeful Look Ahead

After a week in the mountains where we spoke with hardly anyone, we took the train to Padova, and in no time found ourselves in the middle of a bustling scientific meeting. Paolo Bernardi, a man with infectious energy, greeted us with warm hugs, and in the following three days, we were surrounded by scientists from all over the world who are at the forefront of mitochondrial research.

Attending this meeting helped us crystallize and reinforce two bigger points. First, there is a lot that is still unknown about mitochondria, and new findings continue to emerge. Second, it strengthened our belief that the emerging knowledge will open opportunities for treatment. Here's just one example: The focus of the meeting was one enigmatic protein called the mitochondrial permeability transition pore (mitochondrial PTP), which under certain conditions is involved in the death of the cell. Paolo's prominence in science relates to a fascinating finding that he made about this transition pore. Scientists were trying to find this protein for a long time until Paolo provided much evidence that the mitochondrial PTP is in fact ATP synthase, that amazing rotor that is also known as Complex V. In other words, the same rotor that gives us ATP and life can transform under certain conditions to the thing that kills the cell. This concept echoes the revolutionary idea of programmed cell death, which we have described in chapter 1. Recall that in 1996 Xiaodong Wang showed that a protein called cytochrome c, which is an essential part of the electron transport chain that gives us ATP and life, reverses its role when it is released from the mitochondria, and triggers a signaling cascade that tells the cell that it is time to die.

The idea that the life machines can sometimes turn into death machines may sound scary at first, but in fact it is a very hopeful idea. It means that if we understand the pathways leading to the death of the cell, we may be able to develop interventions and therapeutics to prevent their unnecessary death. Case in point, several of the speakers at the meeting described how inhibiting the pore's opening may alleviate cardiovascular and neurodegenerative diseases. As with other meetings about mitochondria that we have attended, there was a sense of hope here, too, that the advances in science may lead to therapeutics.

We already are starting to see how discoveries in mitochondrial biology led to exploring new treatments, and as we argued before, this may lead to a significant shift in medicine that focuses on improving mitochondrial functions and thus our health. Again, we're talking about a shift to a more holistic approach that will not only look at each organ separately but will also recognize that mitochondria can influence our health regardless of symptoms that are apparent in a specific organ. As we emphasized in chapter 5, this approach won't necessarily cure these diseases, but it can reduce the burden on damaged cells and thus tip the balance and allow them to heal. It also may allow other cells to compensate for the loss of injured cells. We come back to our mantra that we are only as healthy as our mitochondria. Having functional mitochondria is the key to our well-being.

HISTORICAL PERSPECTIVE: ON STUBBORNNESS AND GOOD SCIENCE

One reason we're so optimistic about mitochondria's key role in the future of medicine is that we believe that scientific truth has a way of revealing itself, and a lot of it has to do with scientists' stubbornness. Here's an example: The Padova meeting was held in a hall named after Giovanni Felice "Licio" Azzone, who was also a renowned mitochondrial researcher and was Paolo's mentor. In what's known as the oxphos wars, Azzone belonged to the camp that opposed Peter

Mitchell's ideas about how ATP is created. Paolo joined Azzone's lab in 1977, a year before Mitchell won the Nobel Prize, and by then, the dust on this fierce debate was already settled. We once asked Paolo if he ever talked with Azzone about why he and the others in his camp stuck with their theory for so long. He told us that some years after Azzone retired, Paolo reread Azzone's papers from that period and realized to his surprise that these papers contained data that supported Mitchell's hypothesis, but Azzone interpreted these data differently. "I asked him: 'How come you didn't see it? You had the best data to support Mitchell's hypothesis.'" Paolo told us. Azzone listened to him carefully, shook his head, and said: 'You know, I was stubborn'" (*testardo* in Italian).

Stubbornness plays an important role in getting to the truth in science. Azzone and his colleagues were stubborn, but so was Peter Mitchell. At some point during this historic debate, Mitchell printed his ideas in a small gray booklet that he then sent to all his opponents, asking them to conduct experiments that would prove him wrong. They did, and they found a lot of things that were wrong. It is probably true that the stubbornness of his opponents led Peter Mitchell and Jennifer Moyle to refine the theory over the years and develop new experiments to prove it. So while stories about bitter debates often get the headlines, it is this mix of competition and cooperation that often leads to progress. Azzone was among those who received a copy of the gray book from Mitchell, and he gave it to Paolo Bernardi as a gift. "It's one of my most cherished treasures," Paolo told us.

Mitochondria were a hot topic from the 1940s to the 1960s but were considered a relatively sleepy area of research in the following years. Then discoveries by Doug Wallace, Xiaodong Wang, and many other scientists gave a jolt of energy to research in the field. So much so that in March 1999, the journal *Science* dedicated a special issue to the organelle under the title: "Mitochondria Make a Comeback."[5] The new findings generated a renewed interest in the organelle, which apparently did more than just energy transfer, and opened the door to new questions. (Daria still has the cover of this issue of *Science* on a wall in her office.)

Call it stubbornness, persistence, or grit, behind almost every drug development story there is someone who has it, and we have met many stubborn people throughout this book. Just a few examples include Nir Barzilai and his clinical trial for metformin; James McCully, who believes that mitochondrial transplantation provides major health benefits; Carlos Moraes, who hopes to find a solution through mtDNA manipulation. Daria is no stranger to stubbornness either—her mother called her *testa dura,* hard head, the Toscana version of *testardo*—and along with many other scientists, she's been trying to translate discoveries about mitochondria into drugs that will benefit patients. Will this collective persistence have a significant impact on medicine? Only time will tell.

Yet we strongly believe that there are many reasons to be hopeful; again and again academic research shows that by supporting mitochondrial functioning, we support the cell, the tissue, the organ, and ultimately, the whole body. And knowing more about how mitochondria contribute to specific diseases gives us hope that we'll be able to treat diseases more effectively.

Why We Believe in Your Power to Help Your Mitochondria

It isn't just scientific progress in the field that gives us hope. We're also hopeful because as individuals, we have access to unprecedented amounts of health information that can help us if we use it wisely. Despite our many similarities with Ötzi, there is a big difference between us when it comes to how he made his health-related decisions and how we can make our decisions today. Ötzi's use of tattoos as painkillers or his use of birch fungus to treat inflammation were based on beliefs that developed over generations and may or may not have been true. Today, we can base our decisions on scientific evidence. Yet to do this, we need to adopt a systematic way to evaluate information. (See "Think Like a Scientist" in chapter 7.) This is especially true since there is no shortage of advice out there, and while Ötzi might have had to choose between the advice of a couple of local shamans, we are inundated by a never-ending avalanche of advice from healers, doctors, companies, and health gurus on social media. While beliefs are still part of our evaluation, we should try to think like scientists.

Each of us can be hopeful because there's a lot we can do to help our mitochondria, and throughout the book, we outlined evidence-based guidelines for supporting those tiny life machines. Remember that we don't have to cure every mitochondrion. Every action we take to reduce the number of damaged mitochondria in each cell can push it under the threshold and improve the cell's health. Conversely, if we mistreat them, we might push the number of damaged mitochondria above that threshold. What we do can either help our mitochondria or hurt them.

Our genetic makeup plays a role, too, of course. Ötzi had severe arteriosclerosis (hardening of the arteries) even though he presumably lived a very active lifestyle, had no excess fat on his body, and his BMI was just short of 20. Indeed, researchers found that he had a genetic predisposition to cardiovascular diseases. However, while it's true that you can have a disease even when you do everything right, scientists largely believe that lifestyle plays a major role in preventing disease even if you have a genetic predisposition.

When we go for a walk or a run, we are activating a process that may create new life machines and recycle old ones. When we feed our mitochondria with enough food that contains the right nutrients, we make sure our life machines have the fuel they need to operate properly. When we avoid repeatedly overwhelming our mitochondria with too much food, we decrease the likelihood that they will develop metabolic inflexibility. But we should also keep in mind this ancient advice—"Keep to the middle, Icarus"—and not deprive ourselves from essential vitamins and minerals. And by feeding our microbiome with plenty of fiber-rich food and other healthy nutrients, our gut bacteria will return the favor by giving our mitochondria butyrate and other good stuff.

We can also reduce exposure to harmful pollutants; food that was fried in old cooking oil contains toxic aldehydes, and when we avoid them, we may hear our mitochondria breathe a sigh of relief. The same is true for other pollutants: smoking, vaping, alcohol, air pollution, radiation, exhaust fumes, and pesticides. Avoiding prolonged emotional stress is another way to protect our mitochondria. We each have our own way to relax. Whatever it is, when we manage to take a break from our daily stress, we help those tiny organelles.

As it gets dark, when we create the right environment for proper sleep, we'll be helping our mitochondria do a better job tonight. During the daytime, they focused on providing ATP and generating the building blocks our body needs, and now it's time for them to recover and clean up, so we shouldn't overwhelm them with food or alcohol at night.

It's very possible that you've been doing some of this already: exercising, avoiding pollutants, relaxing, sleeping well at night. If that's the case, your 10 million billion mitochondria thank you. But even if you haven't, adopting these healthy practices can turn things around. It's never too late to start. When you adopt these lifestyle changes, you improve the chance that the cell will heal itself, and since mitochondria appear to travel beyond their individual cell, you can harness their power to heal your body as a whole. Are you going to help the life machines help you?

•　•　•

A man is walking in the snow, anxious about what is to be unfolded. It's freezing cold and sunny, and some of the snow from a few days ago has turned into ice. Slippery ice. And then, not far from there, a scream is heard. A cry of pain and hope. A cry of a newborn. The man's phone rings. It's Daria, calling to tell Emanuel about the birth of our granddaughter. A few months earlier our first grandson was born, and some months later, another granddaughter. Each time, a first cry—the first gulp of oxygen from air and not from their mothers' blood. The first handshake between planet Earth and the newborn's life machines. Each time, we marveled (and still do) at this amazing thing called life. In 2022, we spent a winter in Boston, where one of our granddaughters was born. There, on long walks in the snow, this book was born, out of awe of the tiny machines that enable life.

A CONCISE GLOSSARY

**This list includes essential terms that are
most frequently used in the book.**

Aldehydes. A class of molecules that permanently bind to proteins and other molecules and negatively impact their normal functions. We sometimes refer to aldehydes in this book as "rust."

Antioxidants. Molecules that protect cells from the damage caused by oxidants such as free radicals.

Apoptosis. Programmed cell death. A mechanism of controlled killing of cells within a tissue without spilling their contents into the tissue.

ATP. Adenosine triphosphate. This molecule is generated in the *mitochondria* and is used as the energy currency for all the cell's activities.

ATP synthase. Also known as Complex V or ATPase. The rotor spinning inside *mitochondria* generating *ATP*. It is powered by the flow of positively charged *protons*.

Autophagy. Means "self-eating" in Greek; the process of cleaning out damaged components within a cell.

Complex I–Complex IV. Four protein complexes that are embedded in the inner mitochondrial membrane and are part of the *electron transport chain*. As

233

electrons move through this chain of complexes, *protons* are pumped into the intermembrane space, creating the *proton gradient*.

DNA. Deoxyribonucleic acid. The molecule that contains our inherited genes. Nuclear DNA, which we inherit from both our parents, and *mitochondrial DNA,* which we inherit only from our mothers.

Electron transport chain. A process in which electrons are passed from one complex to another, while *protons* move to the intermembrane space, leading to the *proton gradient.* See also *Complex I–Complex IV.*

Fission. Mitochondrial fission is the process by which a mitochondrion is split into two *mitochondria.*

Free radicals. Unstable molecules that are often generated during electron transport between *Complexes I and IV;* their levels increase when the *mitochondria* are impaired. When interacting with *lipids* (which make cells and mitochondrial membranes), they can form toxic *aldehydes.* A class of free radicals are *reactive oxygen species (ROS).*

Fusion. Mitochondrial fusion is the process by which two mitochondria merge.

Krebs cycle. Also called the TCA cycle or Citric Acid cycle. A chain of chemical reactions inside the *mitochondria* that make up the first steps in the process of *ATP* production, as well as the production of some of the cell's building blocks.

Matrix. The inner space of a mitochondrion surrounded by the inner mitochondrial membrane. This is where the *mitochondrial DNA (mtDNA)* is found, and where the *Krebs cycle* takes place.

Metabolite. A general term to describe any substance that is formed or needed for metabolism (the process in which food is broken down and new things are built in our body). For example, glucose is a metabolite of fructose, the sugar found in fruit.

Mitochondria. Organelles essential to life. Responsible for *ATP* generation and many other critical functions. In this book, we refer to them as the life machines. Singular: mitochondrion.

Mitochondrial DNA (mtDNA). The circular *DNA* molecule found in *mitochondria*. There are typically several copies of mtDNA in each mitochondrion.

Mitophagy. The process by which damaged mitochondria are removed, and their parts are "recycled." The *autophagy* of *mitochondria*.

Oxidative phosphorylation (Oxphos). The process of *ATP* production, which includes both the *electron transport chain* and *ATP* synthesis that is done by Complex V (*ATP synthase*).

Oxidative stress. An imbalance between *oxidants* and *antioxidants* in favor of the oxidants. Oxidative stress can cause damage in the cell.

Proton. A subatomic particle with a positive charge, which is found in the center of every element. The protons mentioned in this book are from hydrogen.

Proton gradient. A difference in *proton* concentration between the mitochondrial matrix and the intermembrane space.

Reactive oxygen species (ROS). Types of *free radicals* that contain oxygen and may cause damage in the cell. At a low quantity, they also have useful functions.

ACKNOWLEDGMENTS

Many of the insights and advancements presented here would not have been possible without the work of researchers who have been studying mitochondria for generations. In the same way that one cannot credit a single mitochondrion for giving life to the cell, the work presented in this book is the result of the efforts of thousands of scientists, past and present. We would like to thank each and every one of them. While we obviously couldn't list them all here, nor their important work, we feel that it is important to say that: Thank you! Daria would like to extend special thanks to her lab members and colleagues over the years who have made every day of research such a fun endeavor. She would also like to express her gratitude to the National Institutes of Health and to various foundations for funding her research, as well as the research of other academics; without such public support for basic research, the discovery of new treatments would not be possible.

A big thank-you to all the researchers, practitioners, patients, and their families who shared their insights with us. We are responsible for the accuracy, clarity, and opinions presented on these pages, but we want to express our gratitude to the people who took the time to talk to us. Here, too, we can't list them all, but special thanks to Jen Farmer, Ron Bartek, Elena Ziviani, Melanie Walker, Alicia Kowaltowski, Yair Anikster, Nir Barzilai, Carmen Sandi, Paolo Bernardi, Justin and Erica Sonnenburg, Doug Wallace, Jonathan Brestoff, Kyle Bryant, Pinchas Cohen, Julio Ferreira, Suresh Palaniyandi, Xin Qi, Beatrice Golomb, Richard Haas, Rick Leach, Michael Levitt, David Lynch, James McCully, Carlos Moraes, Robert Naviaux, Martin Picard, Shiladitya Sengupta, Gerald Shadel, Orian Shirihai, William Sly, Elvan Böke, Nadège Zanou, Erin

Flynn-Evans, Anu Suomalainen, Ruth Acton, Mark Tarnopolsky, Natalie Yivgi-Ohana, Christiane Wrann, Jamie Zeitzer, Xiaodong Wang, and Elizabeth Jonas.

Our agent, Jim Levine, supported this project from day one with his insights and guidance. Thanks to Leah Miller at Simon & Schuster for her confidence in our idea. And many thanks to Maria Espinosa, Veronica Alvarado, and the whole Simon Element team for turning this idea into an actual book. We thank Tina Pavlatos for the illustrations, and Reut Diamond for helping with the electron microscopy images. And thanks to Lena Pernas and Shiladitya Sengupta for allowing us to use those images.

We are truly grateful to our friends, family, and colleagues who read early drafts and made valuable comments. Special thanks to Paolo Bernardi, Elisabeth (Zaza) Segre, Kevin Grimes, Betsy Kopmar, Opher Kornfeld, and Aly Elezaby. Last but certainly not least, we would like to thank our extended family, especially our children and their partners, who have been our cheerleaders from day one. For your help and encouragement throughout this project, this book is dedicated to you.

NOTES

Introduction

1. Nick Lane, *Power, Sex, Suicide: Mitochondria and the Meaning of Life*, 2nd ed., Oxford Landmark Science (Oxford, United Kingdom: Oxford University Press, 2018).

2. James D. Shoemaker et al., "Misidentification of Propionic Acid as Ethylene Glycol in a Patient with Methylmalonic Acidemia," *The Journal of Pediatrics* 120, no. 3 (March 1992): 417–21, https://doi.org/10.1016/S0022-3476(05)80909-6; Bill Smith, "Not Guilty. How the System Failed Patricia Stallings," *St. Louis Post-Dispatch*, October 20, 1991; "The Murder That Never Was," People.com, accessed December 12, 2024, https://people.com/archive/the-murder-that-never-was-vol-36-no-23/; "Patricia Stallings—National Registry of Exonerations," accessed December 14, 2024, https://www.law.umich.edu/special/exoneration/Pages/casedetail.aspx?caseid=3660.

Chapter 1:
Yes, the "Powerhouse," but So Much More

1. Hans Adolf Krebs and Anne Martin, *Reminiscences and Reflections* (Oxford: Clarendon Press, 1981).

2. "1988 Nobel Prizes Announced for Physics and for Chemistry," *Nature* 335 (October 27, 1988): 752–53, https://doi.org/10.1038/335752a0.

3. "The Nobel Prize in Physiology or Medicine 1953," NobelPrize.org, accessed December 21, 2024, https://www.nobelprize.org/prizes/medicine/1953/krebs/lecture/.

4. Xuesong Liu et al., "Induction of Apoptotic Program in Cell-Free Extracts: Requirement for dATP and Cytochrome c," *Cell* 86, no. 1 (July 1996): 147–57, https://doi.org/10.1016/S0092-8674(00)80085-9; Guido Kroemer, "Early Work on the Role of Mitochondria in Apoptosis, an Interview with Guido

Kroemer," *Cell Death & Differentiation* 11, no. S1 (July 2004): S33–36, https://doi.org/10.1038/sj.cdd.4401448.

5. Brad Townsend, "Meet the Newly Honored UT Southwestern Biochemist Who Named His Breakthrough Discovery after the Dallas Mavericks," *Dallas Morning News*, October 29, 2018, https://www.dallasnews.com/sports/mavericks/2018/10/29/meet-the-newly-honored-ut-southwestern-biochemist-who-named-his-breakthrough-discovery-after-the-dallas-mavericks/.

6. Lena Pernas and John C. Boothroyd, "Association of Host Mitochondria with the Parasitophorous Vacuole During Toxoplasma Infection Is Not Dependent on Rhoptry Proteins ROP2/8," *International Journal for Parasitology* 40, no. 12 (October 2010): 1367–71, https://doi.org/10.1016/j.ijpara.2010.07.002; Lena Pernas et al., "Mitochondria Restrict Growth of the Intracellular Parasite Toxoplasma Gondii by Limiting Its Uptake of Fatty Acids," *Cell Metabolism* 27, no. 4 (April 2018): 886-897.e4, https://doi.org/10.1016/j.cmet.2018.02.018.

7. Navdeep S. Chandel, *Navigating Metabolism* (Cold Spring Harbor, New York: Cold Spring Harbor Laboratory Press, 2015), xiii.

8. N. S. Chandel et al., "Mitochondrial Reactive Oxygen Species Trigger Hypoxia-Induced Transcription," *Proceedings of the National Academy of Sciences* 95, no. 20 (September 29, 1998): 11715–20, https://doi.org/10.1073/pnas.95.20.11715.

Chapter 2:
On the Move to Help and Heal

1. M. P. Koonce and M. Schliwa, "Bidirectional Organelle Transport Can Occur in Cell Processes That Contain Single Microtubules," *The Journal of Cell Biology* 100, no. 1 (January 1, 1985): 322–26, https://doi.org/10.1083/jcb.100.1.322.

2. Somya Madan et al., "Mitochondria Lead the Way: Mitochondrial Dynamics and Function in Cellular Movements in Development and Disease," *Frontiers in Cell and Developmental Biology* 9 (February 2, 2022): 781933, https://doi.org/10.3389/fcell.2021.781933.

3. The Children's Hospital of Philadelphia, "Facts About Mitochondria," accessed December 12, 2024, https://www.chop.edu/mitochondria-facts.

4. Karin B. Busch, Axel Kowald, and Johannes N. Spelbrink, "Quality Matters: How Does Mitochondrial Network Dynamics and Quality Control Impact on mtDNA Integrity?," *Philosophical Transactions of the Royal Society B: Biological Sciences* 369, no. 1646 (July 5, 2014): 20130442, https://doi.org/10.1098/rstb.2013.0442.

5. Karen Schmitt et al., "Circadian Control of DRP1 Activity Regulates Mito-chondrial Dynamics and Bioenergetics," *Cell Metabolism* 27, no. 3 (March 2018): 657–666.e5, https://doi.org/10.1016/j.cmet.2018.01.011.

6. Albert L. Lehninger, *The Mitochondrion: Molecular Basis of Structure and Function* (New York: W. A. Benjamin, 1964), 31.

7. Qu Tian et al., "Skeletal Muscle Mitochondrial Function Predicts Cognitive Im-pairment and Is Associated with Biomarkers of Alzheimer's Disease and Neu-rodegeneration," *Alzheimer's & Dementia* 19, no. 10 (October 2023): 4436–45, https://doi.org/10.1002/alz.13388; "Dysfunctional Muscle Mitochondria Linked to Higher Dementia Risk," National Institute on Aging, November 9, 2023, https://www.nia.nih.gov/news/dysfunctional-muscle-mitochondria-linked-higher-dementia-risk.

8. Sophia Miliotis et al., "Forms of Extracellular Mitochondria and Their Im-pact in Health," *Mitochondrion* 48 (September 2019): 16–30, https://doi.org/10.1016/j.mito.2019.02.002; Hanbing Li et al., "Transfer and Fates of Dam-aged Mitochondria: Role in Health and Disease," *The FEBS Journal* 291, (March 28, 2024): 5342–64, https://doi.org/10.1111/febs.17119.

9. Chung-ha O. Davis et al., "Transcellular Degradation of Axonal Mitochon-dria," *Proceedings of the National Academy of Sciences* 111, no. 26 (July 2014): 9633–38, https://doi.org/10.1073/pnas.1404651111.

10. Kazuhide Hayakawa et al., "Transfer of Mitochondria from Astrocytes to Neu-rons After Stroke," *Nature* 535, no. 7613 (July 28, 2016): 551–55, https://doi.org/10.1038/nature18928.

11. Martin Pelletier et al., "Platelet Extracellular Vesicles and Their Mitochondrial Content Improve the Mitochondrial Bioenergetics of Cellular Immune Recip-ients," *Transfusion* 63, no. 10 (October 2023): 1983–96, https://doi.org/10.1111/trf.17524; Alexei Grichine et al., "The Fate of Mitochondria During Platelet Activation," *Blood Advances* 7, no. 20 (October 24, 2023): 6290–6302, https://doi.org/10.1182/bloodadvances.2023010423.

12. Nicholas Borcherding et al., "Dietary Lipids Inhibit Mitochondria Transfer to Macrophages to Divert Adipocyte-Derived Mitochondria into the Blood," *Cell Metabolism* 34, no. 10 (October 2022): 1499–1513.e8, https://doi.org/10.1016/j.cmet.2022.08.010.

13. Kitti Thiankhaw et al., "Roles of Humanin and Derivatives on the Pathology of Neurodegenerative Diseases and Cognition," *Biochimica et Biophysica Acta (BBA)—General Subjects* 1866, no. 4 (April 2022): 130097, https://doi.org/10.1016/j.bbagen.2022.130097.

14. Jonathan S. T. Woodhead et al., "High-Intensity Interval Exercise Increases Humanin, a Mitochondrial Encoded Peptide, in the Plasma and Muscle of Men," *Journal of Applied Physiology* 128, no. 5 (May 1, 2020): 1346–54, https://doi.org/10.1152/japplphysiol.00032.2020; Ferdinand Von Walden et al., "Acute Endurance Exercise Stimulates Circulating Levels of Mitochondrial-Derived Peptides in Humans," *Journal of Applied Physiology* 131, no. 3 (September 1, 2021): 1035–42, https://doi.org/10.1152/japplphysiol.00706.2019; Johannes Burtscher et al., "The Muscle-Brain Axis and Neurodegenerative Diseases: The Key Role of Mitochondria in Exercise-Induced Neuroprotection," *International Journal of Molecular Sciences* 22, no. 12 (June 17, 2021): 6479, https://doi.org/10.3390/ijms22126479.

15. Joseph C. Reynolds et al., "MOTS-c Is an Exercise-Induced Mitochondrial-Encoded Regulator of Age-Dependent Physical Decline and Muscle Homeostasis," *Nature Communications* 12, no. 1 (January 20, 2021): 470, https://doi.org/10.1038/s41467-020-20790-0.

16. Noriyuki Fuku et al., "The Mitochondrial-Derived Peptide MOTS-c: A Player in Exceptional Longevity?," *Aging Cell* 14, no. 6 (December 2015): 921–23, https://doi.org/10.1111/acel.12389.

17. Brendan Miller et al., "Mitochondrial DNA Variation in Alzheimer's Disease Reveals a Unique Microprotein Called SHMOOSE," *Molecular Psychiatry* 28, no. 1 (January 2023): 1813–26, https://doi.org/10.1038/s41380-022-01769-3; Hirofumi Zempo et al., "A Pro-Diabetogenic mtDNA Polymorphism in the Mitochondrial-Derived Peptide, MOTS-c," *Aging* 13, no. 2 (January 31, 2021): 1692–1717, https://doi.org/10.18632/aging.202529.

18. Yuichi Hashimoto et al., "A Rescue Factor Abolishing Neuronal Cell Death by a Wide Spectrum of Familial Alzheimer's Disease Genes and Aβ," *Proceedings of the National Academy of Sciences* 98, no. 11 (May 22, 2001): 6336–41, https://doi.org/10.1073/pnas.101133498.

Chapter 3:
The Key to Healthy Aging

1. Pénélope A. Andreux et al., "Mitochondrial Function Is Impaired in the Skeletal Muscle of Pre-Frail Elderly," *Scientific Reports* 8, no. 1 (June 4, 2018): 8548, https://doi.org/10.1038/s41598-018-26944-x.

2. Sabra C. Lewsey et al., "Exercise Intolerance and Rapid Skeletal Muscle Energetic Decline in Human Age-Associated Frailty," *JCI Insight* 5, no. 20 (October 15, 2020): e141246, https://doi.org/10.1172/jci.insight.141246.

3. Kevin R. Short et al., "Decline in Skeletal Muscle Mitochondrial Function with Aging in Humans," *Proceedings of the National Academy of Sciences* 102, no. 15 (April 12, 2005): 5618–23, https://doi.org/10.1073/pnas.0501559102.

4. Graziana Assalve et al., "Exploring the Link Between Telomeres and Mitochondria: Mechanisms and Implications in Different Cell Types," *International Journal of Molecular Sciences* 26, no. 3 (January 24, 2025): 993, https://doi.org/10.3390/ijms26030993.

5. Lynn Sagan, "On the Origin of Mitosing Cells," *Journal of Theoretical Biology* 14, no. 3 (March 1967): 225–74, https://doi.org/10.1016/0022-5193(67)90079-3.

6. Nick Lane, *Power, Sex, Suicide: Mitochondria and the Meaning of Life*, 2nd ed., Oxford Landmark Science (Oxford, United Kingdom: Oxford University Press, 2018).

7. Sara Zapico, "mtDNA Mutations and Their Role in Aging, Diseases and Forensic Sciences," *Aging and Disease* 4, no. 6 (December 1, 2013): 364–80, https://doi.org/10.14336/AD.2013.0400364.

8. Aurélien Riou, Aline Broeglin, and Amandine Grimm, "Mitochondrial Transplantation in Brain Disorders: Achievements, Methods, and Challenges," *Neuroscience & Biobehavioral Reviews* 169 (February 2025): 105971, https://doi.org/10.1016/j.neubiorev.2024.105971. For additional references, see Chapter 5.

9. Margit M. K. Nass and Sylvan Nass, "Intramitochondrial Fibers With DNA Characteristics," *The Journal of Cell Biology* 19, no. 3 (December 1, 1963): 593–611, https://doi.org/10.1083/jcb.19.3.593; William A. Wells, "There's DNA in Those Organelles," *The Journal of Cell Biology* 168, no. 6 (March 14, 2005): 853, https://doi.org/10.1083/jcb1686fta2.

10. R. E. Giles et al., "Maternal Inheritance of Human Mitochondrial DNA," *Proceedings of the National Academy of Sciences* 77, no. 11 (November 1980): 6715–19, https://doi.org/10.1073/pnas.77.11.6715.

11. M. Denaro et al., "Ethnic Variation in Hpa 1 Endonuclease Cleavage Patterns of Human Mitochondrial DNA," *Proceedings of the National Academy of Sciences* 78, no. 9 (September 1981): 5768–72, https://doi.org/10.1073/pnas.78.9.5768.

12. Rebecca L. Cann, Mark Stoneking, and Allan C. Wilson, "Mitochondrial DNA and Human Evolution," *Nature* 325, no. 6099 (January 1987): 31–36, https://doi.org/10.1038/325031a0.

13. Erhan Bilal et al., "Mitochondrial DNA Haplogroup D4a Is a Marker for Extreme Longevity in Japan," ed. Florian Kronenberg, *PLOS One* 3, no. 6 (June 11, 2008): e2421, https://doi.org/10.1371/journal.pone.0002421.

14. For example: Electron Transport Chain, 2017, https://www.youtube.com/watch?v=LQmTKxI4Wn4.

15. Nick Lane, *Power, Sex, Suicide: Mitochondria and the Meaning of Life*, 2nd ed., Oxford Landmark Science (Oxford, United Kingdom: Oxford University Press, 2018).

16. Franklin M. Harold, *To Make the World Intelligible: A Scientist's Journey* (FriesenPress, 2017), 53.

17. John Prebble and Bruce Weber, *Wandering in the Gardens of the Mind: Peter Mitchell and the Making of Glynn* (New York: Oxford University Press, 2003).

18. A proton concentration difference (referred to as membrane potential) between the matrix and the intermembrane space of the mitochondria as a driver of oxidative phosphorylation was reported first by Skulachev and colleagues: E. A. Liberman et al., "Mechanism of Coupling of Oxidative Phosphorylation and the Membrane Potential of Mitochondria," *Nature* 222, no. 5198 (June 1969): 1076–78, https://doi.org/10.1038/2221076a0.

19. For a more detailed definition of oxidative stress, see: Helmut Sies, "Oxidative Stress: Concept and Some Practical Aspects," *Antioxidants* 9, no. 9 (September 10, 2020): 852, https://doi.org/10.3390/antiox9090852.

20. Nir Barzilai and Toni Robino, *Age Later: Health Span, Life Span, and the New Science of Longevity*, 1st ed. (New York: St. Martin's Press, 2020), 19.

21. Kelvin Yen et al., "The Mitochondrial Derived Peptide Humanin Is a Regulator of Lifespan and Healthspan," *Aging* 12, no. 12 (June 23, 2020): 11185–99, https://doi.org/10.18632/aging.103534.

22. Carlos López-Otín et al., "Hallmarks of Aging: An Expanding Universe," *Cell* 186, no. 2 (January 2023): 243–78, https://doi.org/10.1016/j.cell.2022.11.001.

Chapter 4:
Engine Malfunctions

1. J. W. Langston and Jon Palfreman, *The Case of the Frozen Addicts*, 1st ed. (New York: Pantheon Books, 1995).

2. Ashu Johri and M. Flint Beal, "Mitochondrial Dysfunction in Neurodegenerative Diseases," *Journal of Pharmacology and Experimental Therapeutics* 342, no. 3 (September 2012): 619–30, https://doi.org/10.1124/jpet.112.192138; Alexis Diaz-Vegas et al., "Is Mitochondrial Dysfunction a Common Root of Noncommunicable Chronic Diseases?," *Endocrine Reviews* 41, no. 3 (June 1, 2020): bnaa005, https://doi.org/10.1210/endrev/bnaa005.

3. Glenn C. Davis et al., "Chronic Parkinsonism Secondary to Intravenous Injection of Meperidine Analogues," *Psychiatry Research* 1, no. 3 (December 1979): 249–54, https://doi.org/10.1016/0165-1781(79)90006-4.

4. J. William Langston et al., "Chronic Parkinsonism in Humans Due to a Product of Meperidine-Analog Synthesis," *Science* 219, no. 4587 (February 25, 1983): 979–80, https://doi.org/10.1126/science.6823561.

5. Vignayanandam Ravindernath Muddapu et al., "Neurodegenerative Diseases—Is Metabolic Deficiency the Root Cause?," *Frontiers in Neuroscience* 14 (March 31, 2020): 213, https://doi.org/10.3389/fnins.2020.00213.

6. Qixia Wang et al., "Role of Mitophagy in the Neurodegenerative Diseases and Its Pharmacological Advances: A Review," *Frontiers in Molecular Neuroscience* 15 (October 4, 2022): 1014251, https://doi.org/10.3389/fnmol.2022.1014251.

7. Jose E. Galgani, Cedric Moro, and Eric Ravussin, "Metabolic Flexibility and Insulin Resistance," *American Journal of Physiology-Endocrinology and Metabolism* 295, no. 5 (November 2008): E1009–17, https://doi.org/10.1152/ajpendo.90558.2008.

8. CU Boulder Today, University of Colorado Boulder, "Are You Metabolically Flexible? Your New Year's Resolutions May Depend on It," accessed December 12, 2024, https://www.colorado.edu/today/2017/12/22/are-you-metabolically-flexible-your-new-years-resolutions-may-depend-it.

9. Martin Picard and Orian S. Shirihai, "Mitochondrial Signal Transduction," *Cell Metabolism* 34, no. 11 (November 2022): 1620–53, https://doi.org/10.1016/j.cmet.2022.10.008; S. M. Katzman et al., "Mitochondrial Metabolism Reveals a Functional Architecture in Intact Islets of Langerhans from Normal and Diabetic *Psammomys Obesus*," *American Journal of Physiology-Endocrinology and Metabolism* 287, no. 6 (December 2004): E1090–99, https://doi.org/10.1152/ajpendo.00044.2004.

10. "Half of US Adults Have Diabetes or PreDiabetes," *American Institute for Cancer Research* (blog), accessed December 12, 2024, https://www.aicr.org/news/half-of-us-adults-have-diabetes-or-prediabetes/; Andy Menke et al., "Prevalence of and Trends in Diabetes Among Adults in the United States, 1988–2012," *JAMA* 314, no. 10 (September 8, 2015): 1021, https://doi.org/10.1001/jama.2015.10029.

11. Claudia Lewis et al., "Tailoring Exercise Prescription for Effective Diabetes Glucose Management," *The Journal of Clinical Endocrinology & Metabolism* 110, no. Supplement 2 (February 25, 2025): S118–30, https://doi.org/10.1210/clinem/dgae908.

12. The Children's Hospital of Philadelphia, "Facts About Mitochondria," accessed December 12, 2024, https://www.chop.edu/mitochondria-facts.

13. As can be seen in patients of some primary mitochondrial diseases: Deborah E. Meyers, Haseeb Ilias Basha, and Mary Kay Koenig, "Mitochondrial

Cardiomyopathy: Pathophysiology, Diagnosis, and Management," *Texas Heart Institute Journal* 40, no. 4 (2013): 385–94; Rong Tian et al., "Unlocking the Secrets of Mitochondria in the Cardiovascular System: Path to a Cure in Heart Failure—A Report from the 2018 National Heart, Lung, and Blood Institute Workshop," *Circulation* 140, no. 14 (October 2019): 1205–16, https://doi.org/10.1161/CIRCULATIONAHA.119.040551.

14. D. V. Godin, S. Bhimji, and J. H. McNeill, "Effects of Allopurinol Pretreatment on Myocardial Ultrastructure and Arrhythmias Following Coronary Artery Occlusion and Reperfusion," *Virchows Archiv B Cell Pathology Including Molecular Pathology* 52, no. 1 (December 1986): 327–41, https://doi.org/10.1007/BF02889975.

15. "Women Found to Be at Higher Risk for Heart Failure and Heart Attack Death than Men," American Heart Association, accessed December 12, 2024, http://newsroom.heart.org/news/women-found-to-be-at-higher-risk-for-heart-failure-and-heart-attack-death-than-men.

16. A. Eirin, A. Lerman, and L. O. Lerman, "Mitochondrial Injury and Dysfunction in Hypertension-Induced Cardiac Damage," *European Heart Journal* 35, no. 46 (December 2, 2014): 3258–66, https://doi.org/10.1093/eurheartj/ehu436; Gerasimos Siasos et al., "Mitochondria and Cardiovascular Diseases—from Pathophysiology to Treatment," *Annals of Translational Medicine* 6, no. 12 (June 2018): 256, https://doi.org/10.21037/atm.2018.06.21.

17. Doğukan Hazar Ülgen, Silvie Rosalie Ruigrok, and Carmen Sandi, "Powering the Social Brain: Mitochondria in Social Behaviour," *Current Opinion in Neurobiology* 79 (April 2023): 102675, https://doi.org/10.1016/j.conb.2022.102675; Sriparna Ghosal et al., "Mitofusin-2 in Nucleus Accumbens D2-MSNs Regulates Social Dominance and Neuronal Function," *Cell Reports* 42, no. 7 (July 2023): 112776, https://doi.org/10.1016/j.celrep.2023.112776.

18. Na Cai et al., "Molecular Signatures of Major Depression," *Current Biology* 25, no. 9 (May 2015): 1146–56, https://doi.org/10.1016/j.cub.2015.03.008.

19. Teresa E. Daniels, Elizabeth M. Olsen, and Audrey R. Tyrka, "Stress and Psychiatric Disorders: The Role of Mitochondria," *Annual Review of Clinical Psychology* 16, no. 1 (May 7, 2020): 165–86, https://doi.org/10.1146/annurev-clinpsy-082719-104030; Jae Kyung Chung et al., "Investigation of Mitochondrial DNA Copy Number in Patients with Major Depressive Disorder," *Psychiatry Research* 282 (December 2019): 112616, https://doi.org/10.1016/j.psychres.2019.112616.

20. D. Lindqvist et al., "Increased Plasma Levels of Circulating Cell-Free Mitochondrial DNA in Suicide Attempters: Associations with HPA-Axis Hyperac-

tivity," *Translational Psychiatry* 6, no. 12 (December 6, 2016): e971, https://doi .org/10.1038/tp.2016.236; Daniel Lindqvist et al., "Circulating Cell-Free Mitochondrial DNA, but Not Leukocyte Mitochondrial DNA Copy Number, Is Elevated in Major Depressive Disorder," *Neuropsychopharmacology* 43, no. 7 (June 2018): 1557–64, https://doi.org/10.1038/s41386-017-0001-9; Fiona Hollis et al., "Neuroinflammation and Mitochondrial Dysfunction Link Social Stress to Depression," in *Neuroscience of Social Stress*, ed. Klaus A. Miczek and Rajita Sinha, vol. 54, Current Topics in Behavioral Neurosciences (Cham: Springer International Publishing, 2022), 59–93, https://doi.org/10.1007/7854_2021_300.

21. Christopher M. Palmer, *Brain Energy: A Revolutionary Breakthrough in Understanding Mental Health—And Improving Treatment for Anxiety, Depression, OCD, PTSD, and More* (New York: BenBella Books, 2022); Kim Krieger, "Mitochondria Linked to Major Depression in Older Adults," *UConn Today* (blog), February 7, 2023, https://today.uconn.edu/2023/02 /mitochondria-linked-to-major-depression-in-older-adults/; Emory University, Atlanta GA, "Insights into Schizophrenia: Genetic Risk Factor Impairs Mitochondria," accessed December 12, 2024, https://news.emory.edu /stories/2023/08/esc_schizophrenia_genetic_risk_29-08-2023/story.html.

22. Sam Apple, *Ravenous: Otto Warburg, the Nazis, and the Search for the Cancer-Diet Connection*, 1st ed. (New York: Liveright Publishing, 2021).

23. Maria V. Liberti and Jason W. Locasale, "The Warburg Effect: How Does It Benefit Cancer Cells?," *Trends in Biochemical Sciences* 41, no. 3 (March 2016): 211–18, https://doi.org/10.1016/j.tibs.2015.12.001.

24. Tanmoy Saha et al., "Intercellular Nanotubes Mediate Mitochondrial Trafficking Between Cancer and Immune Cells," *Nature Nanotechnology* 17, no. 1 (January 2022): 98–106, https://doi.org/10.1038/s41565-021-01000-4; for additional images (not in cancer), see: "When the Structure of Tunneling Nanotubes (TNTs) Challenges the Very Concept of Cell," Institut Pasteur, January 31, 2019, https://www.pasteur.fr/en/home/research-journal/news/when -structure-tunneling-nanotubes-tnts-challenges-very-concept-cell.

25. Hideki Ikeda et al., "Immune Evasion Through Mitochondrial Transfer in the Tumour Microenvironment," *Nature* 638, no. 8049 (February 6, 2025): 225–36, https://doi.org/10.1038/s41586-024-08439-0.

26. Ilaria Elia and Marcia C. Haigis, "Metabolites and the Tumour Microenvironment: From Cellular Mechanisms to Systemic Metabolism," *Nature Metabolism* 3, no. 1 (January 4, 2021): 21–32, https://doi.org/10.1038/s42255-020-00317-z.

27. Balaraman Kalyanaraman, "Teaching the Basics of Cancer Metabolism: Developing Antitumor Strategies by Exploiting the Differences Between Nor-

mal and Cancer Cell Metabolism," *Redox Biology* 12 (August 2017): 833–42, https://doi.org/10.1016/j.redox.2017.04.018.

28. D. A. Rossignol and R. E. Frye, "Mitochondrial Dysfunction in Autism Spectrum Disorders: A Systematic Review and Meta-Analysis," *Molecular Psychiatry* 17, no. 3 (March 2012): 290–314, https://doi.org/10.1038/mp.2010.136; Cecilia Giulivi et al., "Mitochondrial Dysfunction in Autism," *JAMA* 304, no. 21 (December 1, 2010): 2389, https://doi.org/10.1001/jama.2010.1706.

29. "Meet the 'Mitomaniacs' Who Say Mitochondria Matter in Autism," The Transmitter: Neuroscience News and Perspectives, November 22, 2021, https://www.thetransmitter.org/spectrum/meet-the-mitomaniacs-who-say-mitochondria-matter-in-autism/.

30. Tal Yardeni et al., "An mtDNA Mutant Mouse Demonstrates That Mitochondrial Deficiency Can Result in Autism Endophenotypes," *Proceedings of the National Academy of Sciences* 118, no. 6 (February 9, 2021): e2021429118, https://doi.org/10.1073/pnas.2021429118.

31. Robert K. Naviaux, "Metabolic Features of the Cell Danger Response," *Mitochondrion* 16 (May 2014): 7–17, https://doi.org/10.1016/j.mito.2013.08.006.

32. Robert K. Naviaux et al., "Low-Dose Suramin in Autism Spectrum Disorder: A Small, Phase I/II, Randomized Clinical Trial," *Annals of Clinical and Translational Neurology* 4, no. 7 (July 2017): 491–505, https://doi.org/10.1002/acn3.424; "Researchers Studying Century-Old Drug in Potential New Approach to Autism," accessed December 12, 2024, https://today.ucsd.edu/story/researchers_studying_century_old_drug_in_potential_new_approach_to_autism.

33. David Hough et al., "Randomized Clinical Trial of Low Dose Suramin Intravenous Infusions for Treatment of Autism Spectrum Disorder," *Annals of General Psychiatry* 22, no. 1 (November 6, 2023): 45, https://doi.org/10.1186/s12991-023-00477-8.

34. Richard E. Frye, "Mitochondrial Dysfunction in Autism Spectrum Disorder: Unique Abnormalities and Targeted Treatments," *Seminars in Pediatric Neurology* 35 (October 2020): 100829, https://doi.org/10.1016/j.spen.2020.100829.

35. Brian Vastag, "She Wrote to a Scientist About Her Fatigue. It Inspired a Breakthrough," *Washington Post*, September 17, 2023, https://www.washingtonpost.com/health/2023/09/17/fatigue-cfs-longcovid-mitochondria/; Ping-yuan Wang et al., "WASF3 Disrupts Mitochondrial Respiration and May Mediate Exercise Intolerance in Myalgic Encephalomyelitis/Chronic Fatigue Syndrome," *Proceedings of the National Academy of Sciences* 120, no. 34 (August 22, 2023): e2302738120, https://doi.org/10.1073/pnas.2302738120.

36. Brent Appelman et al., "Muscle Abnormalities Worsen After Post-Exertional Malaise in Long COVID," *Nature Communications* 15, no. 1 (January 4, 2024): 17, https://doi.org/10.1038/s41467-023-44432-3; Tihamer Molnar et al., "Mitochondrial Dysfunction in Long COVID: Mechanisms, Consequences, and Potential Therapeutic Approaches," *GeroScience* 46, no. 5 (April 26, 2024): 5267–86, https://doi.org/10.1007/s11357-024-01165-5.

37. Esther De Boer et al., "Decreased Fatty Acid Oxidation and Altered Lactate Production During Exercise in Patients with Post-Acute COVID-19 Syndrome," *American Journal of Respiratory and Critical Care Medicine* 205, no. 1 (January 1, 2022): 126–29, https://doi.org/10.1164/rccm.202108-1903LE.

38. Emily Wood, Katherine H. Hall, and Warren Tate, "Role of Mitochondria, Oxidative Stress and the Response to Antioxidants in Myalgic Encephalomyelitis/Chronic Fatigue Syndrome: A Possible Approach to SARS-CoV-2 'Long-haulers'?," *Chronic Diseases and Translational Medicine* 7, no. 1 (March 2021): 14–26, https://doi.org/10.1016/j.cdtm.2020.11.002.

39. Laetitia Gay et al., "Long-Term Persistence of Mitochondrial Dysfunctions After Viral Infections and Antiviral Therapies: A Review of Mechanisms Involved," *Journal of Medical Virology* 96, no. 9 (September 2024): e29886, https://doi.org/10.1002/jmv.29886.

40. Bindu D. Paul et al., "Redox Imbalance Links COVID-19 and Myalgic Encephalomyelitis/Chronic Fatigue Syndrome," *Proceedings of the National Academy of Sciences* 118, no. 34 (August 24, 2021): e2024358118, https://doi.org/10.1073/pnas.2024358118.

41. Aida Rodríguez-Nuevo et al., "Oocytes Maintain ROS-Free Mitochondrial Metabolism by Suppressing Complex I," *Nature* 607, no. 7920 (July 28, 2022): 756–61, https://doi.org/10.1038/s41586-022-04979-5.

42. For example, Sumit Parikh et al., "Diagnosis and Management of Mitochondrial Disease: A Consensus Statement from the Mitochondrial Medicine Society," *Genetics in Medicine* 17, no. 9 (September 2015): 689–701, https://doi.org/10.1038/gim.2014.177.

43. Shi En Kim, "To Study Aging, Scientists Are Looking to Outer Space," *National Geographic*, December 2, 2020, https://www.nationalgeographic.com/science/article/to-study-aging-scientists-are-looking-to-outer-space-iss.

44. "Mitochondrial Changes Key to Health Problems in Space—NASA," November 25, 2020, https://www.nasa.gov/humans-in-space/mitochondrial-changes-key-to-health-problems-in-space/; Willian A. Da Silveira et al., "Comprehensive Multi-Omics Analysis Reveals Mitochondrial Stress as a Central Biologi-

cal Hub for Spaceflight Impact," *Cell* 183, no. 5 (November 2020): 1185–1201. e20, https://doi.org/10.1016/j.cell.2020.11.002.

45. Martin Picard and Orian S. Shirihai, "Mitochondrial Signal Transduction," *Cell Metabolism* 34, no. 11 (November 2022): 1620–53, https://doi.org/10.1016 /j.cmet.2022.10.008.

Chapter 5:
The Upside of Tending to Our Mitochondria

1. Vincent Paupe et al., "Impaired Nuclear Nrf2 Translocation Undermines the Oxidative Stress Response in Friedreich Ataxia," ed. Antoni L. Andreu, *PLOS One* 4, no. 1 (January 22, 2009): e4253, https://doi.org/10.1371/journal .pone.0004253.

2. Sylvia Boesch and Elisabetta Indelicato, "Approval of Omaveloxolone for Friedreich Ataxia," *Nature Reviews Neurology* 20, no. 6 (June 2024): 313–14, https://doi.org/10.1038/s41582-024-00957-9.

3. Cody J. Schmidlin et al., "Redox Regulation by NRF2 in Aging and Disease," *Free Radical Biology and Medicine* 134 (April 2019): 702–7, https://doi.org /10.1016/j.freeradbiomed.2019.01.016.

4. C. A. Bannister et al., "Can People with Type 2 Diabetes Live Longer than Those Without? A Comparison of Mortality in People Initiated with Metformin or Sulphonylurea Monotherapy and Matched, Non-diabetic Controls," *Diabetes, Obesity and Metabolism* 16, no. 11 (November 2014): 1165–73, https://doi .org/10.1111/dom.12354; Jared M. Campbell et al., "Metformin Reduces All-Cause Mortality and Diseases of Ageing Independent of Its Effect on Diabetes Control: A Systematic Review and Meta-Analysis," *Ageing Research Reviews* 40 (November 2017): 31–44, https://doi.org/10.1016/j.arr.2017.08.003; Willy Marcos Valencia et al., "Metformin and Ageing: Improving Ageing Outcomes Beyond Glycaemic Control," *Diabetologia* 60, no. 9 (September 2017): 1630–38, https://doi.org/10.1007/s00125-017-4349-5.

5. Chris R. Triggle et al., "Metformin: Is It a Drug for All Reasons and Diseases?," *Metabolism* 133 (August 2022): 155223, https://doi.org/10.1016 /j.metabol.2022.155223; Isabella Backman, "Metformin: How a Widely Used Diabetes Medication Actually Works," accessed December 12, 2024, https:// medicine.yale.edu/news-article/how-a-widely-used-diabetes-medication -actually-works/.

6. Also see: Gretchen Reynolds, "An Anti-Aging Pill? Think Twice," *New York Times*, June 19, 2019, https://www.nytimes.com/2019/06/19/well/move/an-anti

-aging-pill-think-twice.html; "Metformin (Oral Route)," Mayo Clinic, accessed April 23, 2025, https://www.mayoclinic.org/drugs-supplements/metformin -oral-route/description/drg-20067074.

7. Carlos López-Otín et al., "Hallmarks of Aging: An Expanding Universe," *Cell* 186, no. 2 (January 2023): 243–78, https://doi.org/10.1016/j.cell.2022.11.001.

8. Nir Barzilai and Toni Robino, *Age Later: Health Span, Life Span, and the New Science of Longevity*, 1st ed. (New York: St. Martin's Press, 2020), 15–16.

9. Stephen S. Hall, "The Man Who Wants to Beat Back Aging," *Science*, September 16, 2015, https://www.science.org/content/article/feature-man-who -wants-beat-back-aging.

10. Xin Qi et al., "A Novel Drp1 Inhibitor Diminishes Aberrant Mitochondrial Fission and Neurotoxicity," *Journal of Cell Science* 126, no. 3 (February 2013): 789–802, https://doi.org/10.1242/jcs.114439.

11. Xing Guo et al., "Inhibition of Mitochondrial Fragmentation Diminishes Huntington's Disease–Associated Neurodegeneration," *Journal of Clinical Investigation* 123, no. 12 (December 2, 2013): 5371–88, https://doi.org/10.1172 /JCI70911.

12. Ankita Srivastava et al., "Therapeutic Potential of P110 Peptide: New Insights into Treatment of Alzheimer's Disease," *Life* 13, no. 11 (November 2, 2023): 2156, https://doi.org/10.3390/life13112156; Emily Filichia et al., "Inhibition of Drp1 Mitochondrial Translocation Provides Neural Protection in Dopaminergic System in a Parkinson's Disease Model Induced by MPTP," *Scientific Reports* 6, no. 1 (September 13, 2016): 32656, https://doi.org/10.1038/srep32656; Amit U Joshi et al., "Inhibition of Drp1/Fis1 Interaction Slows Progression of Amyotrophic Lateral Sclerosis," *EMBO Molecular Medicine* 10, no. 3 (March 2018): e8166, https://doi.org/10.15252/emmm.201708166.

13. Fucheng Luo et al., "Inhibition of Drp1 Hyper-Activation Is Protective in Animal Models of Experimental Multiple Sclerosis," *Experimental Neurology* 292 (June 2017): 21–34, https://doi.org/10.1016/j.expneurol.2017.02.015; Zhenyu Chen et al., "Inhibiting Mitochondrial Fission Rescues Degeneration in Hereditary Spastic Paraplegia Neurons," *Brain* 145, no. 11 (November 21, 2022): 4016–31, https://doi.org/10.1093/brain/awab488; Xing Guo, Hiromi Sesaki, and Xin Qi, "Drp1 Stabilizes P53 on the Mitochondria to Trigger Necrosis Under Oxidative Stress Conditions *in Vitro* and *in Vivo*," *Biochemical Journal* 461, no. 1 (July 1, 2014): 137–46, https://doi.org/10.1042/BJ20131438; Nicole L. Mancini et al., "Perturbed Mitochondrial Dynamics Is a Novel Feature of Colitis That Can Be Targeted to Lessen Disease," *Cellular and Molecular Gastroenterology and Hepatology* 10, no. 2 (2020): 287–307, https://doi.org/10.1016

/j.jcmgh.2020.04.004; Preethy S. Sridharan et al., "Acutely Blocking Excessive Mitochondrial Fission Prevents Chronic Neurodegeneration After Traumatic Brain Injury," *Cell Reports Medicine* 5, no. 9 (September 2024): 101715, https://doi.org/10.1016/j.xcrm.2024.101715; Haikel Dridi et al., "Aberrant Mitochondrial Dynamics Contributes to Diaphragmatic Weakness Induced by Mechanical Ventilation," ed. Andrey Abramov, *PNAS Nexus* 2, no. 11 (November 1, 2023): pgad336, https://doi.org/10.1093/pnasnexus/pgad336.

14. P. Hemachandra Reddy, "Inhibitors of Mitochondrial Fission as a Therapeutic Strategy for Diseases with Oxidative Stress and Mitochondrial Dysfunction," *Journal of Alzheimer's Disease* 40, no. 2 (March 31, 2014): 245–56, https://doi.org/10.3233/JAD-132060.

15. Raz Bar-Ziv et al., "Glial-Derived Mitochondrial Signals Affect Neuronal Proteostasis and Aging," *Science Advances* 9, no. 41 (October 13, 2023): eadi1411, https://doi.org/10.1126/sciadv.adi1411; Mitch Leslie, "Faulty Communication Between Organs Could Make Us Old," accessed December 18, 2024, https://www.science.org/content/article/faulty-communication-organs-make-us-old.

16. Eyad Katrangi et al., "Xenogenic Transfer of Isolated Murine Mitochondria into Human ρ^0 Cells Can Improve Respiratory Function," *Rejuvenation Research* 10, no. 4 (December 2007): 561–70, https://doi.org/10.1089/rej.2007.0575.

17. Nancy Fliesler, "How Mitochondrial Transfer Restores Heart Muscle," Boston Children's Answers, May 21, 2024, https://answers.childrenshospital.org/mitochondrial-transfer/.

18. Pedro Norat et al., "Intraarterial Transplantation of Mitochondria After Ischemic Stroke Reduces Cerebral Infarction," *Stroke: Vascular and Interventional Neurology* 3, no. 3 (May 2023): e000644, https://doi.org/10.1161/SVIN.122.000644; Melanie Walker et al., "Autologous Mitochondrial Transplant for Acute Cerebral Ischemia: Phase 1 Trial Results and Review," *Journal of Cerebral Blood Flow & Metabolism*, (December 4, 2024): 0271678X241305230, https://doi.org/10.1177/0271678X241305230.

19. Caryn M. Cloer et al., "Mitochondrial Transplant After Ischemia Reperfusion Promotes Cellular Salvage and Improves Lung Function During Ex-Vivo Lung Perfusion," *The Journal of Heart and Lung Transplantation* 42, no. 5 (May 2023): 575–84, https://doi.org/10.1016/j.healun.2023.01.002; Aybuke Celik et al., "Mitochondrial Transplantation: Effects on Chemotherapy in Prostate and Ovarian Cancer Cells *in Vitro* and *in Vivo*," *Biomedicine & Pharmacotherapy* 161 (May 2023): 114524, https://doi.org/10.1016/j.biopha.2023.114524; Alfredo Cruz-Gregorio et al., "Mitochondrial Transplantation Strategies in Multifaceted Induction of Cancer Cell Death," *Life Sciences* 332 (November

2023): 122098, https://doi.org/10.1016/j.lfs.2023.122098; Jui-Chih Chang et al., "Mitochondrial Transplantation Regulates Antitumour Activity, Chemoresistance and Mitochondrial Dynamics in Breast Cancer," *Journal of Experimental & Clinical Cancer Research* 38, no. 1 (December 2019): 30, https://doi.org/10.1186/s13046-019-1028-z; James D. McCully, Pedro J. Del Nido, and Sitaram M. Emani, "Mitochondrial Transplantation: The Advance to Therapeutic Application and Molecular Modulation," *Frontiers in Cardiovascular Medicine* 10 (December 15, 2023): 1268814, https://doi.org/10.3389/fcvm.2023.1268814; Tasnim Arroum et al., "Mitochondrial Transplantation's Role in Rodent Skeletal Muscle Bioenergetics: Recharging the Engine of Aging," *Biomolecules* 14, no. 4 (April 18, 2024): 493, https://doi.org/10.3390/biom14040493; Elad Jacoby et al., "Mitochondrial Augmentation of Hematopoietic Stem Cells in Children with Single Large-Scale Mitochondrial DNA Deletion Syndromes," *Science Translational Medicine* 14, no. 676 (December 21, 2022): eabo3724, https://doi.org/10.1126/scitranslmed.abo3724; Ritsuko Nakai et al., "Mitochondria Transfer-Based Therapies Reduce the Morbidity and Mortality of Leigh Syndrome," *Nature Metabolism* 6, no. 10 (September 2, 2024): 1886–96, https://doi.org/10.1038/s42255-024-01125-5.

20. Megan Molteni, "In a First, Children with Rare Genetic Diseases Get Mitochondrial Transplants from Their Mothers," *STAT* (blog), December 21, 2022, https://www.statnews.com/2022/12/21/children-with-rare-genetic-disease-get-mitochondrial-transplant-from-mothers/; Elad Jacoby et al., "Mitochondrial Augmentation of Hematopoietic Stem Cells in Children with Single Large Scale Mitochondrial DNA Deletion Syndromes," *Science Translational Medicine* 14, no. 676 (December 21, 2022): eabo3724, https://doi.org/10.1126/scitranslmed.abo3724.

21. Ritsuko Nakai et al., "Mitochondria Transfer-Based Therapies Reduce the Morbidity and Mortality of Leigh Syndrome," *Nature Metabolism* 6, no. 10 (September 2, 2024): 1886–96, https://doi.org/10.1038/s42255-024-01125-5.

22. Andrea L. Gropman and Alexis C. Komor, "Personalized Gene Editing to Treat an Inborn Error of Metabolism," *New England Journal of Medicine*, May 15, 2025, NEJMe2505721, https://doi.org/10.1056/NEJMe2505721.

23. Danielle Gerhard, PhD, "On the Road to Treating Mitochondrial Disease," Drug Discover News, Development & Diagnostics Articles | DDN Magazine, accessed April 23, 2025, https://www.drugdiscoverynews.com/on-the-road-to-treating-mitochondrial-disease-15334; Wendy K. Shoop et al., "Efficient Elimination of MELAS-Associated m.3243G Mutant Mitochondrial DNA by an Engineered mitoARCUS Nuclease," *Nature Me-*

tabolism 5, no. 12 (November 30, 2023): 2169–83, https://doi.org/10.1038/s42255-023-00932-6.

24. Zilong Bian et al., "Genetic Predisposition, Modifiable Lifestyles, and Their Joint Effects on Human Lifespan: Evidence from Multiple Cohort Studies," *BMJ Evidence-Based Medicine* 29, no. 4 (August 2024): 255–63, https://doi.org/10.1136/bmjebm-2023-112583.

Chapter 6:
Go, Kyle, Go! Mitochondria and Exercise

1. James M. Hagberg et al., "The Historical Context and Scientific Legacy of John O. Holloszy," *Journal of Applied Physiology* 127, no. 2 (August 1, 2019): 277–305, https://doi.org/10.1152/japplphysiol.00669.2018; Juleen R. Zierath, James M. Hagberg, and Edward F. Coyle, "John O. Holloszy (1933–2018)," *Cell Metabolism* 28, no. 3 (September 2018): 329, https://doi.org/10.1016/j.cmet.2018.08.016; *ACSM's Distinguished Leaders—John Holloszy, M.D., FACSM*, 2022, https://www.youtube.com/watch?v=m0UTX-v1b9E.

2. J. O. Holloszy, "Biochemical Adaptations in Muscle. Effects of Exercise on Mitochondrial Oxygen Uptake and Respiratory Enzyme Activity in Skeletal Muscle," *The Journal of Biological Chemistry* 242, no. 9 (May 10, 1967): 2278–82.

3. Shanyao Zhou et al., "Running Improves Muscle Mass by Activating Autophagic Flux and Inhibiting Ubiquitination Degradation in Mdx Mice," *Gene* 899 (March 2024): 148136, https://doi.org/10.1016/j.gene.2024.148136.

4. Carolina Simioni et al., "Oxidative Stress: Role of Physical Exercise and Antioxidant Nutraceuticals in Adulthood and Aging," *Oncotarget* 9, no. 24 (March 30, 2018): 17181–98, https://doi.org/10.18632/oncotarget.24729.

5. Johannes Burtscher et al., "The Muscle-Brain Axis and Neurodegenerative Diseases: The Key Role of Mitochondria in Exercise-Induced Neuroprotection," *International Journal of Molecular Sciences* 22, no. 12 (June 17, 2021): 6479, https://doi.org/10.3390/ijms22126479.

6. Jens Frey Halling and Henriette Pilegaard, "PGC-1α-Mediated Regulation of Mitochondrial Function and Physiological Implications," *Applied Physiology, Nutrition, and Metabolism* 45, no. 9 (September 2020): 927–36, https://doi.org/10.1139/apnm-2020-0005.

7. Pontus Boström et al., "A PGC1-α-Dependent Myokine That Drives Brown-Fat-like Development of White Fat and Thermogenesis," *Nature* 481, no. 7382 (January 2012): 463–68, https://doi.org/10.1038/nature10777.

8. Johannes Burtscher et al., "The Muscle-Brain Axis and Neurodegenerative Diseases: The Key Role of Mitochondria in Exercise-Induced Neuroprotection," *International Journal of Molecular Sciences* 22, no. 12 (June 17, 2021): 6479, https://doi.org/10.3390/ijms22126479.

9. Mohammad R. Islam et al., "Exercise Hormone Irisin Is a Critical Regulator of Cognitive Function," *Nature Metabolism* 3, no. 8 (August 20, 2021): 1058–70, https://doi.org/10.1038/s42255-021-00438-z.

10. Steffen Maak et al., "Progress and Challenges in the Biology of FNDC5 and Irisin," *Endocrine Reviews* 42, no. 4 (July 16, 2021): 436–56, https://doi.org/10.1210/endrev/bnab003; Elke Albrecht et al., "Irisin: Still Chasing Shadows," *Molecular Metabolism* 34 (April 2020): 124–35, https://doi.org/10.1016/j.molmet.2020.01.016.

11. Tsubasa Tomoto and Rong Zhang, "Arterial Aging and Cerebrovascular Function: Impact of Aerobic Exercise Training in Older Adults," *Aging and Disease* 15, no. 4 (2024): 1672–78, https://doi.org/10.14336/AD.2023.1109-1.

12. Matt Kaufman et al., "The Role of Physical Exercise in Cognitive Preservation: A Systematic Review," *American Journal of Lifestyle Medicine* 18, no. 4 (July 2024): 574–91, https://doi.org/10.1177/15598276231201555.

13. Nazik Elgaddal, Ellen A. Kramarow, and Cynthia Reuben, "Physical Activity Among Adults Aged 18 and Over: United States, 2020" (National Center for Health Statistics, August 30, 2022), https://doi.org/10.15620/cdc:120213; U.S. Department of Health and Human Services, "Physical Activity Guidelines for Americans, 2nd ed.," 2018, https://odphp.health.gov/paguidelines/second-edition/pdf/Physical_Activity_Guidelines_2nd_edition.pdf.

14. Daniel Lieberman, *Exercised: Why Something We Never Evolved to Do Is Healthy and Rewarding* (New York: Pantheon Books, 2020), 256.

15. Kyle Bryant and Alex Schnitzler, *Shifting into High Gear* (Deerfield Beach, FL: Health Communications, 2019).

16. Kyle Bryant and Alex Schnitzler, *Shifting into High Gear* (Deerfield Beach, FL: Health Communications, 2019).

17. Mikel Izquierdo et al., "Global Consensus on Optimal Exercise Recommendations for Enhancing Healthy Longevity in Older Adults (ICFSR)," *The Journal of Nutrition, Health and Aging* 29, no. 1 (January 2025): 100401, https://doi.org/10.1016/j.jnha.2024.100401.

18. "How Much Exercise Do You Really Need?," Mayo Clinic, accessed December 12, 2024, https://www.mayoclinic.org/healthy-lifestyle/fitness/expert-answers/exercise/faq-20057916.

19. "How to Calculate Heart Rate Zones," Cleveland Clinic, accessed December 12, 2024, https://health.clevelandclinic.org/exercise-heart-rate-zones-explained.

20. Ian R. Lanza and K. Sreekumaran Nair, "Mitochondrial Metabolic Function Assessed *in Vivo* and *in Vitro*," *Current Opinion in Clinical Nutrition and Metabolic Care* 13, no. 5 (September 2010): 511–17, https://doi.org/10.1097/MCO.0b013e32833cc93d.

21. For example, "Everything to Know About VO$_2$ Max," Healthline, April 22, 2020, https://www.healthline.com/health/vo2-max; "VO$_2$ Max," in Wikipedia, October 6, 2024, https://en.wikipedia.org/w/index.php?title=VO2_max&oldid=1249741971.

22. An Ouyang et al., "Effects of Aerobic Exercise on Brain Age and Health in Middle-Aged and Older Adults: A Single-Arm Pilot Clinical Trial," *Life* 14, no. 7 (July 8, 2024): 855, https://doi.org/10.3390/life14070855.

23. Matthew M. Robinson et al., "Enhanced Protein Translation Underlies Improved Metabolic and Physical Adaptations to Different Exercise Training Modes in Young and Old Humans," *Cell Metabolism* 25, no. 3 (March 2017): 581–92, https://doi.org/10.1016/j.cmet.2017.02.009.

24. Brian A. Irving et al., "Combined Training Enhances Skeletal Muscle Mitochondrial Oxidative Capacity Independent of Age," *The Journal of Clinical Endocrinology & Metabolism* 100, no. 4 (April 1, 2015): 1654–63, https://doi.org/10.1210/jc.2014-3081.

25. Ferdinand Von Walden et al., "Acute Endurance Exercise Stimulates Circulating Levels of Mitochondrial-Derived Peptides in Humans," *Journal of Applied Physiology* 131, no. 3 (September 1, 2021): 1035–42, https://doi.org/10.1152/japplphysiol.00706.2019.

26. Joseph C. Reynolds et al., "MOTS-c Is an Exercise-Induced Mitochondrial-Encoded Regulator of Age-Dependent Physical Decline and Muscle Homeostasis," *Nature Communications* 12, no. 1 (January 20, 2021): 470, https://doi.org/10.1038/s41467-020-20790-0; Changhan Lee, Kyung Hwa Kim, and Pinchas Cohen, "MOTS-c: A Novel Mitochondrial-Derived Peptide Regulating Muscle and Fat Metabolism," *Free Radical Biology and Medicine* 100 (November 2016): 182–87, https://doi.org/10.1016/j.freeradbiomed.2016.05.015.

27. Carrie Arnold, "After Obesity Drugs' Success, Companies Rush to Preserve Skeletal Muscle," *Nature Biotechnology* 42, no. 3 (March 2024): 351–53, https://doi.org/10.1038/s41587-024-02176-5.

28. Anna-Maria Joseph, Peter J. Adhihetty, and Christiaan Leeuwenburgh, "Beneficial Effects of Exercise on Age-Related Mitochondrial Dysfunction and Oxidative Stress in Skeletal Muscle," *The Journal of Physiology* 594, no. 18 (September

15, 2016): 5105–23, https://doi.org/10.1113/JP270659; Charles Affourtit and Jane E. Carré, "Mitochondrial Involvement in Sarcopenia," *Acta Physiologica* 240, no. 3 (March 2024): e14107, https://doi.org/10.1111/apha.14107.

29. Mikael Flockhart et al., "Excessive Exercise Training Causes Mitochondrial Functional Impairment and Decreases Glucose Tolerance in Healthy Volunteers," *Cell Metabolism* 33, no. 5 (May 2021): 957–70.e6, https://doi.org/10.1016/j.cmet.2021.02.017.

30. Shlomit Chevion et al., "Plasma Antioxidant Status and Cell Injury After Severe Physical Exercise," *Proceedings of the National Academy of Sciences* 100, no. 9 (April 29, 2003): 5119–23, https://doi.org/10.1073/pnas.0831097100.

31. Thomas H. Julian et al., "Physical Exercise Is a Risk Factor for Amyotrophic Lateral Sclerosis: Convergent Evidence from Mendelian Randomisation, Transcriptomics and Risk Genotypes," *EBioMedicine* 68 (June 2021): 103397, https://doi.org/10.1016/j.ebiom.2021.103397.

32. Carly S. Cox et al., "Mitohormesis in Mice via Sustained Basal Activation of Mitochondrial and Antioxidant Signaling," *Cell Metabolism* 28, no. 5 (November 2018): 776–786.e5, https://doi.org/10.1016/j.cmet.2018.07.011.

33. Troy L. Merry and Michael Ristow, "Mitohormesis in Exercise Training," *Free Radical Biology and Medicine* 98 (September 2016): 123–30, https://doi.org/10.1016/j.freeradbiomed.2015.11.032.

34. Estelle Balan et al., "Endurance Training Alleviates MCP-1 and TERRA Accumulation at Old Age in Human Skeletal Muscle," *Experimental Gerontology* 153 (October 2021): 111510, https://doi.org/10.1016/j.exger.2021.111510.

35. M. Fiorenza et al., "Hormetic Modulation of Angiogenic Factors by Exercise-Induced Mechanical and Metabolic Stress in Human Skeletal Muscle," *American Journal of Physiology-Heart and Circulatory Physiology* 319, no. 4 (October 1, 2020): H824–34, https://doi.org/10.1152/ajpheart.00432.2020.

36. Adam C. Jordan, Christopher G. R. Perry, and Arthur J. Cheng, "Promoting a Pro-Oxidant State in Skeletal Muscle: Potential Dietary, Environmental, and Exercise Interventions for Enhancing Endurance-Training Adaptations," *Free Radical Biology and Medicine* 176 (November 2021): 189–202, https://doi.org/10.1016/j.freeradbiomed.2021.09.014.

37. Estelle Balan et al., "Regular Endurance Exercise Promotes Fission, Mitophagy, and Oxidative Phosphorylation in Human Skeletal Muscle Independently of Age," *Frontiers in Physiology* 10 (August 22, 2019): 1088, https://doi.org/10.3389/fphys.2019.01088.

38. Yoan Arribat et al., "Distinct Patterns of Skeletal Muscle Mitochondria Fusion, Fission and Mitophagy upon Duration of Exercise Training," *Acta Phys-*

iologica 225, no. 2 (February 2019): e13179, https://doi.org/10.1111/apha.13179.

39. Carrie Arnold, "After Obesity Drugs' Success, Companies Rush to Preserve Skeletal Muscle," *Nature Biotechnology* 42, no. 3 (March 2024): 351–53, https://doi.org/10.1038/s41587-024-02176-5.

40. Gavriela Voulgaridou et al., "Increasing Muscle Mass in Elders Through Diet and Exercise: A Literature Review of Recent RCTs," *Foods* 12, no. 6 (March 13, 2023): 1218, https://doi.org/10.3390/foods12061218.

41. Katie Schütze et al., "Old Muscle, New Tricks: A Clinician Perspective on Sarcopenia and Where to Next," *Current Opinion in Neurology* 36, no. 5 (October 2023): 441–49, https://doi.org/10.1097/WCO.0000000000001185.

42. Daniel H. Aslan, Joshua M. Collette, and Justus D. Ortega, "Bicycling Exercise Helps Maintain a Youthful Metabolic Cost of Walking in Older Adults," *Journal of Aging and Physical Activity* 29, no. 1 (February 1, 2021): 36–42, https://doi.org/10.1123/japa.2019-0327.

43. Henan Zhao et al., "Long-Term Voluntary Running Prevents the Onset of Symptomatic Friedreich's Ataxia in Mice," *Scientific Reports* 10, no. 1 (April 8, 2020): 6095, https://doi.org/10.1038/s41598-020-62952-6.

44. Andreas Mæchel Fritzen et al., "Effect of Aerobic Exercise Training and Deconditioning on Oxidative Capacity and Muscle Mitochondrial Enzyme Machinery in Young and Elderly Individuals," *Journal of Clinical Medicine* 9, no. 10 (September 26, 2020): 3113, https://doi.org/10.3390/jcm9103113; Andreas M. Fritzen et al., "Adaptations in Mitochondrial Enzymatic Activity Occurs Independent of Genomic Dosage in Response to Aerobic Exercise Training and Deconditioning in Human Skeletal Muscle," *Cells* 8, no. 3 (March 12, 2019): 237, https://doi.org/10.3390/cells8030237.

Chapter 7:
Feeding the Life Machines

1. Ovid, *Metamorphoses*, trans. Stephanie McCarter (New York: Penguin Books, 2022), 216–17.

2. Alicia Kowaltowski and Fernando Abdulkader, *Where Does All That Food Go? How Metabolism Fuels Life* (Cham: Springer International Publishing AG, 2020).

3. Food Insight, "2018 Food and Health Survey," *Food Insight* (blog), January 13, 2019, https://foodinsight.org/2018-food-and-health-survey/.

4. About ultra-processed food: They are "high in unhealthy types of fat, refined starches, free sugars and salt, and poor sources of protein, dietary fibre and

micronutrients." Carlos Augusto Monteiro et al., "The UN Decade of Nutrition, the NOVA Food Classification and the Trouble with Ultra-Processing," *Public Health Nutrition* 21, no. 1 (January 2018): 5–17, https://doi.org/10.1017/S1368980017000234.

5. Deborah M. Muoio, "Metabolic Inflexibility: When Mitochondrial Indecision Leads to Metabolic Gridlock," *Cell* 159, no. 6 (December 2014): 1253–62, https://doi.org/10.1016/j.cell.2014.11.034; Sheena Faherty, "How Overeating May Contribute to a Metabolic 'Traffic Jam,'" *Scientific American*, accessed December 12, 2024, https://www.scientificamerican.com/blog/guest-blog/how-overeating-may-contribute-to-a-metabolic-traffic-jam/; Jose Serrano et al., "Effect of Dietary Bioactive Compounds on Mitochondrial and Metabolic Flexibility," *Diseases* 4, no. 1 (March 10, 2016): 14, https://doi.org/10.3390/diseases4010014.

6. Alicia Kowaltowski and Fernando Abdulkader, *Where Does All That Food Go? How Metabolism Fuels Life* (Cham: Springer International Publishing AG, 2020), 92–94; On mitochondria and obesity: Juan C. Bournat and Chester W. Brown, "Mitochondrial Dysfunction in Obesity," *Current Opinion in Endocrinology, Diabetes and Obesity* 17, no. 5 (October 2010): 446–52, https://doi.org/10.1097/MED.0b013e32833c3026.

7. Grishma Hirode and Robert J. Wong, "Trends in the Prevalence of Metabolic Syndrome in the United States, 2011–2016," *JAMA* 323, no. 24 (June 23, 2020): 2526, https://doi.org/10.1001/jama.2020.4501; Pankaj Prasun, "Mitochondrial Dysfunction in Metabolic Syndrome," *Biochimica et Biophysica Acta (BBA)—Molecular Basis of Disease* 1866, no. 10 (October 2020): 165838, https://doi.org/10.1016/j.bbadis.2020.165838.

8. Ian R. Lanza et al., "Chronic Caloric Restriction Preserves Mitochondrial Function in Senescence Without Increasing Mitochondrial Biogenesis," *Cell Metabolism* 16, no. 6 (December 2012): 777–88, https://doi.org/10.1016/j.cmet.2012.11.003; G. López-Lluch et al., "Calorie Restriction Induces Mitochondrial Biogenesis and Bioenergetic Efficiency," *Proceedings of the National Academy of Sciences* 103, no. 6 (February 7, 2006): 1768–73, https://doi.org/10.1073/pnas.0510452103; Chad R. Hancock et al., "Does Calorie Restriction Induce Mitochondrial Biogenesis? A Reevaluation," *The FASEB Journal* 25, no. 2 (February 2011): 785–91, https://doi.org/10.1096/fj.10-170415.

9. U.S. Department of Agriculture and U.S. Department of Health and Human Services. *Dietary Guidelines for Americans, 2020-2025*, 9th ed. (December 2020). https://www.dietaryguidelines.gov/.

10. Aleksandra Kaliszewska et al., "The Interaction of Diet and Mitochondrial Dysfunction in Aging and Cognition," *International Journal of Molecular Sciences* 22, no. 7 (March 30, 2021): 3574, https://doi.org/10.3390/ijms22073574.; Fernanda M. Cerqueira and Alicia J. Kowaltowski, "Commonly Adopted Caloric Restriction Protocols Often Involve Malnutrition," *Ageing Research Reviews* 9, no. 4 (October 2010): 424–30, https://doi.org/10.1016/j.arr.2010.05.002; Ricarda Schmidt et al., "Macro- and Micronutrient Intake in Children with Avoidant/Restrictive Food Intake Disorder," *Nutrients* 13, no. 2 (January 27, 2021): 400, https://doi.org/10.3390/nu13020400.

11. Gary Fanjiang and Ronald E Kleinman, "Nutrition and Performance in Children," *Current Opinion in Clinical Nutrition and Metabolic Care* 10, no. 3 (May 2007): 342–47, https://doi.org/10.1097/MCO.0b013e3280523a9e; Akanksha Likhar, Prerna Baghel, and Manoj Patil, "Early Childhood Development and Social Determinants," *Cureus* 14, no. 9 (September 23, 2022): e29500, https://doi.org/10.7759/cureus.29500.

12. Victor M. Victor et al., "Altered Mitochondrial Function and Oxidative Stress in Leukocytes of Anorexia Nervosa Patients," ed. Siyaram Pandey, *PLOS One* 9, no. 9 (September 25, 2014): e106463, https://doi.org/10.1371/journal.pone.0106463.

13. Fernanda M. Cerqueira and Alicia J. Kowaltowski, "Commonly Adopted Caloric Restriction Protocols Often Involve Malnutrition," *Ageing Research Reviews* 9, no. 4 (October 2010): 424–30, https://doi.org/10.1016/j.arr.2010.05.002.

14. "The Nobel Prize in Physiology or Medicine 1929," NobelPrize.org, accessed December 12, 2024, https://www.nobelprize.org/prizes/medicine/1929/eijkman/lecture/.

15. "Thiamin Deficiency—Disorders of Nutrition," Merck Manual Consumer Version, accessed December 12, 2024, https://www.merckmanuals.com/home/disorders-of-nutrition/vitamins/thiamin-deficiency.

16. Joëlle J. E. Janssen et al., "Mito-Nuclear Communication by Mitochondrial Metabolites and Its Regulation by B-Vitamins," *Frontiers in Physiology* 10 (February 12, 2019): 78, https://doi.org/10.3389/fphys.2019.00078.

17. "What's the Beef with Red Meat?," Harvard Health, February 1, 2020, https://www.health.harvard.edu/staying-healthy/whats-the-beef-with-red-meat; Maryam S. Farvid et al., "Consumption of Red Meat and Processed Meat and Cancer Incidence: A Systematic Review and Meta-Analysis of Prospective Studies," *European Journal of Epidemiology* 36, no. 9 (September 2021): 937–51, https://doi.org/10.1007/s10654-021-00741-9; Xiao Gu et al., "Red Meat Intake and Risk of Type 2 Diabetes in a Prospective Cohort Study of United

States Females and Males," *The American Journal of Clinical Nutrition* 118, no. 6 (December 2023): 1153–63, https://doi.org/10.1016/j.ajcnut.2023.08.021; Wenming Shi et al., "Red Meat Consumption, Cardiovascular Diseases, and Diabetes: A Systematic Review and Meta-Analysis," *European Heart Journal* 44, no. 28 (July 21, 2023): 2626–35, https://doi.org/10.1093/eurheartj/ehad336.

18. Christopher M. Palmer, *Brain Energy: A Revolutionary Breakthrough in Understanding Mental Health—And Improving Treatment for Anxiety, Depression, OCD, PTSD, and More* (New York: BenBella Books, 2022), 234.

19. Mount Sinai, New York, "Vitamin C (Ascorbic Acid) Information," Mount Sinai Health System, accessed December 12, 2024, https://www.mountsinai .org/health-library/supplement/vitamin-c-ascorbic-acid.

20. David W. Killilea and Alison N. Killilea, "Mineral Requirements for Mitochondrial Function: A Connection to Redox Balance and Cellular Differentiation," *Free Radical Biology and Medicine* 182 (March 2022): 182–91, https://doi.org /10.1016/j.freeradbiomed.2022.02.022.

21. Man Liu et al., "Magnesium Supplementation Improves Diabetic Mitochondrial and Cardiac Diastolic Function," *JCI Insight* 4, no. 1 (January 10, 2019): e123182, https://doi.org/10.1172/jci.insight.123182.

22. Javier S. Perona, "Membrane Lipid Alterations in the Metabolic Syndrome and the Role of Dietary Oils," *Biochimica et Biophysica Acta (BBA)—Biomembranes* 1859, no. 9 (September 2017): 1690–1703, https://doi.org/10.1016 /j.bbamem.2017.04.015.

23. "9 High Fat Foods That Are Actually Super Healthy," Healthline, June 4, 2017, https://www.healthline.com/nutrition/10-super-healthy-high-fat-foods; "Polyunsaturated Fats," www.heart.org, accessed December 12, 2024, https://www .heart.org/en/healthy-living/healthy-eating/eat-smart/fats/polyunsaturated-fats.

24. James J. DiNicolantonio and James O'Keefe, "The Importance of Maintaining a Low Omega-6/Omega-3 Ratio for Reducing the Risk of Autoimmune Diseases, Asthma, and Allergies," *Missouri Medicine* 118, no. 5 (2021): 453–59.

25. WebMD Editorial Contributor, "Healthy Foods High in Omega-6," WebMD, accessed December 16, 2024, https://www.webmd.com/diet/foods-high-in-omega-6.

26. Victor Hoffbrand, *The Folate Story: A Vitamin Under the Microscope* (San Jose: Matador, 2023).

27. Mount Sinai, New York, "Vitamin B9 (Folic Acid) Information," Mount Sinai Health System, accessed December 12, 2024, https://www.mountsinai.org /health-library/supplement/vitamin-b9-folic-acid; "Office of Dietary Supplements—Folate," accessed March 18, 2025, https://ods.od.nih.gov/factsheets /Folate-HealthProfessional/.

28. Michaela E. Murphy and Cara J. Westmark, "Folic Acid Fortification and Neural Tube Defect Risk: Analysis of the Food Fortification Initiative Dataset," *Nutrients* 12, no. 1 (January 18, 2020): 247, https://doi.org/10.3390/nu12010247; not everyone agrees with this analysis, as can be seen in a follow-up comment on this article: Vijaya Kancherla et al., "The Fallacy of Using Administrative Data in Assessing the Effectiveness of Food Fortification. Comment on: 'Folic Acid Fortification and Neural Tube Defect Risk: Analysis of the Food Fortification Initiative Dataset. Nutrients 2020, 12, 247,'" *Nutrients* 12, no. 5 (May 8, 2020): 1352, https://doi.org/10.3390/nu12051352.

29. "Effects of Lowering Homocysteine Levels with B Vitamins on Cardiovascular Disease, Cancer, and Cause-Specific Mortality Meta-Analysis of 8 Randomized Trials Involving 37,485 Individuals Effects of Lowering Homocysteine Levels," *Archives of Internal Medicine* 170, no. 18 (October 11, 2010): 1622, https://doi.org/10.1001/archinternmed.2010.348; "Office of Dietary Supplements—Folate," accessed March 18, 2025, https://ods.od.nih.gov/fact sheets/Folate-HealthProfessional/.

30. Mount Sinai, New York, "Multiple Vitamin Overdose Information," Mount Sinai Health System, accessed December 12, 2024, https://www.mountsinai .org/health-library/poison/multiple-vitamin-overdose; For additional discussion of supplements, see: Alice Callahan, "Should I Be Taking Supplements?," *New York Times*, October 31, 2023, https://www.nytimes.com/2023/10/31/well /eat/supplements-health-benefits.html.

31. Melissa Vos et al., "Vitamin K_2 Is a Mitochondrial Electron Carrier That Rescues Pink1 Deficiency," *Science* 336, no. 6086 (June 8, 2012): 1306–10, https:// doi.org/10.1126/science.1218632.

32. "Why Vitamin K Can Be Dangerous If You Take Warfarin," Cleveland Clinic, accessed April 23, 2025, https://health.clevelandclinic.org/vitamin-k-can -dangerous-take-warfarin.

33. Eija Pirinen et al., "Niacin Cures Systemic NAD+ Deficiency and Improves Muscle Performance in Adult-Onset Mitochondrial Myopathy," *Cell Metabolism* 31, no. 6 (June 2020): 1078–1090.e5, https://doi.org/10.1016/j.cmet .2020.04.008.

34. Lok-Kin Yeung et al., "Multivitamin Supplementation Improves Memory in Older Adults: A Randomized Clinical Trial," *The American Journal of Clinical Nutrition* 118, no. 1 (July 2023): 273–82, https://doi.org/10.1016 /j.ajcnut.2023.05.011; also see: Chirag M. Vyas et al., "Effect of Multivitamin-Mineral Supplementation versus Placebo on Cognitive Function: Results from the Clinic Subcohort of the COcoa Supplement and Multivitamin Outcomes

Study (COSMOS) Randomized Clinical Trial and Meta-Analysis of 3 Cognitive Studies Within COSMOS," *The American Journal of Clinical Nutrition* 119, no. 3 (March 2024): 692–701, https://doi.org/10.1016/j.ajcnut.2023.12.011.

35. Francine Grodstein et al., "Long-Term Multivitamin Supplementation and Cognitive Function in Men: A Randomized Trial," *Annals of Internal Medicine* 159, no. 12 (December 17, 2013): 806–14, https://doi.org/10.7326/0003-4819 -159-12-201312170-00006.

36. Council for Responsible Nutrition, "Three-Quarters of Americans Take Dietary Supplements; Most Users Agree They Are Essential to Maintaining Health, CRN Consumer Survey Finds," accessed December 12, 2024, https://www.crnusa .org/newsroom/three-quarters-americans-take-dietary-supplements-most -users-agree-they-are-essential; "What Supplements Do You Need? Probably None," Columbia University Irving Medical Center, January 24, 2023, https:// www.cuimc.columbia.edu/news/what-supplements-do-you-need-probably -none; Do You Need a Daily Supplement?," Harvard Health, September 1, 2018, https://www.health.harvard.edu/staying-healthy/do-you-need-a-daily -supplement; "Should You Take Dietary Supplements?," NIH News in Health, accessed December 12, 2024, https://newsinhealth.nih.gov/2013/08/should -you-take-dietary-supplements.

37. "The Nutrition Source," *The Nutrition Source* (blog), accessed December 12, 2024, https://nutritionsource.hsph.harvard.edu/.

38. Michael Ristow et al., "Antioxidants Prevent Health-Promoting Effects of Physical Exercise in Humans," *Proceedings of the National Academy of Sciences* 106, no. 21 (May 26, 2009): 8665–70, https://doi.org/10.1073/pnas.0903485106.

39. "Antioxidants," *The Nutrition Source* (blog), September 18, 2012, https:// nutritionsource.hsph.harvard.edu/antioxidants/.

40. Marie E. Migaud, Mathias Ziegler, and Joseph A. Baur, "Regulation of and Challenges in Targeting NAD+ Metabolism," *Nature Reviews Molecular Cell Biology* 25, no. 10 (October 2024): 822–40, https://doi.org/10.1038/s41580-024-00752-w.

41. Beatrice A. Golomb et al., "Coenzyme Q10 Benefits Symptoms in Gulf War Veterans: Results of a Randomized Double-Blind Study," *Neural Computation* 26, no. 11 (November 2014): 2594–2651, https://doi.org/10.1162 /NECO_a_00659; Beatrice A. Golomb et al., "Mitochondrial Impairment but Not Peripheral Inflammation Predicts Greater Gulf War Illness Severity," *Scientific Reports* 13, no. 1 (July 12, 2023): 10739, https://doi.org/10.1038/s41598 -023-35896-w.

42. Panagiota Florou et al., "Does Coenzyme Q10 Supplementation Improve Fertility Outcomes in Women Undergoing Assisted Reproductive Technology Pro-

cedures? A Systematic Review and Meta-Analysis of Randomized-Controlled Trials," *Journal of Assisted Reproduction and Genetics* 37, no. 10 (October 2020): 2377–87, https://doi.org/10.1007/s10815-020-01906-3.

43. Svend A. Mortensen et al., "The Effect of Coenzyme Q10 on Morbidity and Mortality in Chronic Heart Failure," *JACC: Heart Failure* 2, no. 6 (December 2014): 641–49, https://doi.org/10.1016/j.jchf.2014.06.008.

44. Hua Qu et al., "Effects of Coenzyme Q10 on Statin-Induced Myopathy: An Updated Meta-Analysis of Randomized Controlled Trials," *Journal of the American Heart Association* 7, no. 19 (October 2, 2018): e009835, https://doi .org/10.1161/JAHA.118.009835.

45. "Coenzyme Q10," Mayo Clinic, accessed December 12, 2024, https://www .mayoclinic.org/drugs-supplements-coenzyme-q10/art-20362602.

46. Meghan J. Ho, Edmond Ck Li, and James M. Wright, "Blood Pressure Lowering Efficacy of Coenzyme Q10 for Primary Hypertension," ed. Cochrane Hypertension Group, *Cochrane Database of Systematic Reviews* 2016, no. 3 (March 3, 2016), https://doi.org/10.1002/14651858.CD007435.pub3.

47. "Analysis: Some Natural Supplements Can Be Dangerously Contaminated," PBS News, February 19, 2020, https://www.pbs.org/newshour/health /analysis-some-natural-supplements-can-be-dangerously-contaminated; C. Michael White, "Dietary Supplements Pose Real Dangers to Patients," *Annals of Pharmacotherapy* 54, no. 8 (August 2020): 815–19, https://doi.org/10.1177 /1060028019900504.

48. Office of the Commissioner, "Weight Loss, Male Enhancement and Other Products Sold Online or in Stores May Be Dangerous," *FDA*, December 8, 2021, https://www.fda.gov/consumers/consumer-updates/weight-loss-male -enhancement-and-other-products-sold-online-or-stores-may-be-dangerous.

49. Lauren A. E. Erland and Praveen K. Saxena, "Melatonin Natural Health Products and Supplements: Presence of Serotonin and Significant Variability of Melatonin Content," *Journal of Clinical Sleep Medicine* 13, no. 2 (February 15, 2017): 275–81, https://doi.org/10.5664/jcsm.6462.

50. Johann Grundlingh et al., "2,4-Dinitrophenol (DNP): A Weight Loss Agent with Significant Acute Toxicity and Risk of Death," *Journal of Medical Toxicology* 7, no. 3 (September 2011): 205–12, https://doi.org/10.1007/s13181-011 -0162-6.

51. Evelyn B. Parr, Brooke L. Devlin, and John A. Hawley, "Perspective: Time-Restricted Eating—Integrating the What with the When," *Advances in Nutrition* 13, no. 3 (May 2022): 699–711, https://doi.org/10.1093/advances/nmac015.

52. Heather J. Weir et al., "Dietary Restriction and AMPK Increase Lifespan via Mitochondrial Network and Peroxisome Remodeling," *Cell Metabolism* 26, no. 6 (December 2017): 884–896.e5, https://doi.org/10.1016/j.cmet.2017.09.024.

53. Andrea Di Francesco et al., "Dietary Restriction Impacts Health and Lifespan of Genetically Diverse Mice," *Nature* 634, no. 8034 (October 17, 2024): 684–92, https://doi.org/10.1038/s41586-024-08026-3; Elie Dolgin, "Eating Less Can Lead to a Longer Life: Massive Study in Mice Shows Why," *Nature*, (October 9, 2024): d41586-024-03277-6, https://doi.org/10.1038/d41586-024-03277-6; Nicholas J. Schork, "Dietary Restriction Can Extend Lifespan—but Genetics Matters More," *Nature* 634, no. 8034 (October 17, 2024): 555–56, https://doi.org/10.1038/d41586-024-03055-4.

54. Marie-Pierre St-Onge et al., "Meal Timing and Frequency: Implications for Cardiovascular Disease Prevention: A Scientific Statement from the American Heart Association," *Circulation* 135, no. 9 (February 28, 2017): e96-e121, https://doi.org/10.1161/CIR.0000000000000476.

55. Richard P. Evershed et al., "Dairying, Diseases and the Evolution of Lactase Persistence in Europe," *Nature* 608, no. 7922 (August 11, 2022): 336–45, https://doi.org/10.1038/s41586-022-05010-7.

56. "MLA Top Health Websites," MLA, accessed December 12, 2024, https://www.mlanet.org/resources/information-for-patients-and-caregivers/mla-top-health-websites/.

Chapter 8:
You, Mitochondria, and Your Gut

1. "What Is Your Gut Microbiome?," Cleveland Clinic, accessed December 12, 2024, https://my.clevelandclinic.org/health/body/25201-gut-microbiome; Zhijie Wan et al., "Intermediate Role of Gut Microbiota in Vitamin B Nutrition and Its Influences on Human Health," *Frontiers in Nutrition* 9 (December 13, 2022): 1031502, https://doi.org/10.3389/fnut.2022.1031502.

2. Aaron G. Wexler et al., "Human Gut Bacteroides Capture Vitamin B_{12} via Cell Surface-Exposed Lipoproteins," *eLife* 7 (September 18, 2018): e37138, https://doi.org/10.7554/eLife.37138.

3. Megan W. Bourassa et al., "Butyrate, Neuroepigenetics and the Gut Microbiome: Can a High Fiber Diet Improve Brain Health?," *Neuroscience Letters* 625 (June 2016): 56–63, https://doi.org/10.1016/j.neulet.2016.02.009; "What

Is Butyrate and What Can It Do?," Cleveland Clinic, accessed December 13, 2024, https://health.clevelandclinic.org/butyrate-benefits.

4. Audrey Rivière et al., "Bifidobacteria and Butyrate-Producing Colon Bacteria: Importance and Strategies for Their Stimulation in the Human Gut," *Frontiers in Microbiology* 7 (June 28, 2016), https://doi.org/10.3389/fmicb.2016.00979.

5. Sara Daniela Gomes et al., "The Role of Diet Related Short-Chain Fatty Acids in Colorectal Cancer Metabolism and Survival: Prevention and Therapeutic Implications," *Current Medicinal Chemistry* 27, no. 24 (July 7, 2020): 4087–4108, https://doi.org/10.2174/0929867325666180530102050; Xin Li et al., "Sodium Butyrate Ameliorates Oxidative Stress-Induced Intestinal Epithelium Barrier Injury and Mitochondrial Damage Through AMPK-Mitophagy Pathway," ed. Hao Wu, *Oxidative Medicine and Cellular Longevity* 2022 (January 29, 2022): 3745135, https://doi.org/10.1155/2022/3745135.

6. Andrew Reynolds et al., "Carbohydrate Quality and Human Health: A Series of Systematic Reviews and Meta-Analyses," *The Lancet* 393, no. 10170 (February 2019): 434–45, https://doi.org/10.1016/S0140-6736(18)31809-9.

7. Probiotics, Prebiotics and Postbiotics. The differences here: https://health.clevelandclinic.org/postbiotics.

8. Lisette Voytko-Best, "Maye Musk—Elon's Mom—Plugs $100 Anti-Aging Supplement, But Says She Hasn't 'Noticed' if It's Working," Forbes, accessed April 24, 2025, https://www.forbes.com/sites/lisettevoytko/2021/09/28/maye-musk-elons-mom-plugs-100-anti-aging-supplement-but-says-i-havent-noticed-if-its-working/; Dongryeol Ryu et al., "Urolithin A Induces Mitophagy and Prolongs Lifespan in *C. Elegans* and Increases Muscle Function in Rodents," *Nature Medicine* 22, no. 8 (August 2016): 879–88, https://doi.org/10.1038/nm.4132; Sophia Liu et al., "Effect of Urolithin A Supplementation on Muscle Endurance and Mitochondrial Health in Older Adults: A Randomized Clinical Trial," *JAMA Network Open* 5, no. 1 (January 20, 2022): e2144279, https://doi.org/10.1001/jamanetworkopen.2021.44279; A. Cortés-Martín et al., "The Gut Microbiota Urolithin Metabotypes Revisited: The Human Metabolism of Ellagic Acid Is Mainly Determined by Aging," *Food & Function* 9, no. 8 (2018): 4100–4106, https://doi.org/10.1039/C8FO00956B.

9. Bingnan Liu et al., "Gut Microbiota Regulates Host Melatonin Production Through Epithelial Cell MyD88," *Gut Microbes* 16, no. 1 (December 31, 2024): 2313769, https://doi.org/10.1080/19490976.2024.2313769.

10. Maren Gesper et al., "Gut-Derived Metabolite Indole-3-Propionic Acid Modulates Mitochondrial Function in Cardiomyocytes and Alters Cardiac

Function," *Frontiers in Medicine* 8 (March 22, 2021): 648259, https://doi.org /10.3389/fmed.2021.648259.

11. Justin Sonnenburg and Erica Sonnenburg, *The Good Gut: Taking Control of Your Weight, Your Mood, and Your Long-Term Health* (New York: Penguin Books, 2015), 18; Frances Spragge et al., "Microbiome Diversity Protects Against Pathogens by Nutrient Blocking," *Science* 382, no. 6676 (December 15, 2023): eadj3502, https://doi.org/10.1126/science.adj3502.

12. Dennis R. Ownby, "Exposure to Dogs and Cats in the First Year of Life and Risk of Allergic Sensitization at 6 to 7 Years of Age," *JAMA* 288, no. 8 (August 28, 2002): 963, https://doi.org/10.1001/jama.288.8.963.

13. Justin Sonnenburg and Erica Sonnenburg, *The Good Gut: Taking Control of Your Weight, Your Mood, and Your Long-Term Health* (New York: Penguin Books, 2015), 46; Sivan Kijner, Oren Kolodny, and Moran Yassour, "Human Milk Oligosaccharides and the Infant Gut Microbiome from an Eco-Evolutionary Perspective," *Current Opinion in Microbiology* 68 (August 2022): 102156, https:// doi.org/10.1016/j.mib.2022.102156.

14. Kristen A. Earle et al., "Quantitative Imaging of Gut Microbiota Spatial Organization," *Cell Host & Microbe* 18, no. 4 (October 2015): 478–88, https://doi .org/10.1016/j.chom.2015.09.002.

15. Yiming Zhang, Jindong Zhang, and Liping Duan, "The Role of Microbiota-Mitochondria Crosstalk in Pathogenesis and Therapy of Intestinal Diseases," *Pharmacological Research* 186 (December 2022): 106530, https://doi.org /10.1016/j.phrs.2022.106530.

16. Hannah C. Wastyk et al., "Gut-Microbiota-Targeted Diets Modulate Human Immune Status," *Cell* 184, no. 16 (August 2021): 4137–4153.e14, https://doi .org/10.1016/j.cell.2021.06.019.

17. Fivos Borbolis, Eirini Mytilinaiou, and Konstantinos Palikaras, "The Crosstalk Between Microbiome and Mitochondrial Homeostasis in Neurodegeneration," *Cells* 12, no. 3 (January 28, 2023): 429, https://doi.org/10.3390/cells12030429.

18. Vanessa O. Zambelli et al., "Aldehyde Dehydrogenase-2 Regulates Nociception in Rodent Models of Acute Inflammatory Pain," *Science Translational Medicine* 6, no. 251 (August 27, 2014): 251ra118, https://doi.org/10.1126 /scitranslmed.3009539.

19. Chiara Morreale et al., "Microbiota and Pain: Save Your Gut Feeling," *Cells* 11, no. 6 (March 11, 2022): 971, https://doi.org/10.3390/cells11060971.

20. Barbara Fyntanidou et al., "Probiotics in Postoperative Pain Management," *Journal of Personalized Medicine* 13, no. 12 (November 25, 2023): 1645, https:// doi.org/10.3390/jpm13121645.

21. David Brenner et al., "Pain After Upper Limb Surgery Under Peripheral Nerve Block Is Associated with Gut Microbiome Composition and Diversity," *Neurobiology of Pain* 10 (August 2021): 100072, https://doi.org/10.1016/j.ynpai.2021.100072.

22. Arianna K. DeGruttola et al., "Current Understanding of Dysbiosis in Disease in Human and Animal Models," *Inflammatory Bowel Diseases* 22, no. 5 (May 2016): 1137–50, https://doi.org/10.1097/MIB.0000000000000750; Kannayiram Alagiakrishnan, Joao Morgadinho, and Tyler Halverson, "Approach to the Diagnosis and Management of Dysbiosis," *Frontiers in Nutrition* 11 (April 19, 2024): 1330903, https://doi.org/10.3389/fnut.2024.1330903.

23. "Food Sources of Dietary Fiber—Dietary Guidelines for Americans," accessed December 17, 2024, https://www.dietaryguidelines.gov/resources/2020-2025-dietary-guidelines-online-materials/food-sources-select-nutrients/food-sources-fiber; "How Much Fiber Is Found in Common Foods?," Mayo Clinic, accessed December 17, 2024, https://www.mayoclinic.org/healthy-lifestyle/nutrition-and-healthy-eating/in-depth/high-fiber-foods/art-20050948.

24. Mount Sinai, New York, "Soluble vs. Insoluble Fiber Information," Mount Sinai Health System, accessed December 15, 2024, https://www.mountsinai.org/health-library/special-topic/soluble-vs-insoluble-fiber.

25. "Antibiotics Can Temporarily Wipe out the Gut Microbiome | UCLA Health," accessed April 23, 2025, https://www.uclahealth.org/news/article/antibiotics-can-temporarily-wipe-out-gut-microbiome.

26. Jotham Suez et al., "Post-Antibiotic Gut Mucosal Microbiome Reconstitution Is Impaired by Probiotics and Improved by Autologous FMT," *Cell* 174, no. 6 (September 2018): 1406–1423.e16, https://doi.org/10.1016/j.cell.2018.08.047.

27. Binhui Pan et al., "A Meta-Analysis of Microbial Therapy Against Metabolic Syndrome: Evidence from Randomized Controlled Trials," *Frontiers in Nutrition* 8 (December 15, 2021): 775216, https://doi.org/10.3389/fnut.2021.775216.

Chapter 9:
Give Them a Break

1. Zachary A. Caddick et al., "A Review of the Environmental Parameters Necessary for an Optimal Sleep Environment," *Building and Environment* 132 (March 2018): 11–20, https://doi.org/10.1016/j.buildenv.2018.01.020.

2. Carolin Franziska Reichert, Tom Deboer, and Hans-Peter Landolt, "Adenosine, Caffeine, and Sleep–Wake Regulation: State of the Science and Perspectives," *Journal of Sleep Research* 31, no. 4 (August 2022): e13597, https://doi.org/10.1111

/jsr.13597; "Adenosine and Sleep: Understanding Your Sleep Drive," Sleep Foundation, June 7, 2022, https://www.sleepfoundation.org/how-sleep-works/adenosine-and-sleep.

3. "Circadian Rhythms," National Institute of General Medical Sciences (NIGMS), accessed December 13, 2024, https://www.nigms.nih.gov/education/fact-sheets/Pages/circadian-rhythms.aspx.

4. Anouk Charlot et al., "Beneficial Effects of Early Time-Restricted Feeding on Metabolic Diseases: Importance of Aligning Food Habits with the Circadian Clock," *Nutrients* 13, no. 5 (April 22, 2021): 1405, https://doi.org/10.3390/nu13051405.

5. Yaarit Adamovich et al., "Circadian Clocks and Feeding Time Regulate the Oscillations and Levels of Hepatic Triglycerides," *Cell Metabolism* 19, no. 2 (February 2014): 319–30, https://doi.org/10.1016/j.cmet.2013.12.016.

6. Hans Reinke and Gad Asher, "Circadian Clock Control of Liver Metabolic Functions," *Gastroenterology* 150, no. 3 (March 2016): 574–80, https://doi.org/10.1053/j.gastro.2015.11.043; Nityanand Bolshette et al., "Circadian Regulation of Liver Function: From Molecular Mechanisms to Disease Pathophysiology," *Nature Reviews Gastroenterology & Hepatology* 20, no. 11 (November 2023): 695–707, https://doi.org/10.1038/s41575-023-00792-1.

7. Lisa L. Morselli, Aurore Guyon, and Karine Spiegel, "Sleep and Metabolic Function," *Pflügers Archiv—European Journal of Physiology* 463, no. 1 (January 2012): 139–60, https://doi.org/10.1007/s00424-011-1053-z.

8. Simeng Zhang et al., "Metabolic Flexibility During Sleep," *Scientific Reports* 11, no. 1 (September 8, 2021): 17849, https://doi.org/10.1038/s41598-021-97301-8.

9. Gad Asher and Paolo Sassone-Corsi, "Time for Food: The Intimate Interplay between Nutrition, Metabolism, and the Circadian Clock," *Cell* 161, no. 1 (March 2015): 84–92, https://doi.org/10.1016/j.cell.2015.03.015.

10. Irshaad O. Ebrahim et al., "Alcohol and Sleep I: Effects on Normal Sleep," *Alcoholism: Clinical and Experimental Research* 37, no. 4 (April 2013): 539–49, https://doi.org/10.1111/acer.12006.

11. Li-Feng Jiang-Xie et al., "Neuronal Dynamics Direct Cerebrospinal Fluid Perfusion and Brain Clearance," *Nature* 627, no. 8002 (March 7, 2024): 157–64, https://doi.org/10.1038/s41586-024-07108-6.

12. William Gibson, *Pattern Recognition* (New York: Berkley Books, 2004).

13. *From Toads and Sheep to Chronotherapy: A Melatonin Story // Josephine Arendt*, 2018, https://www.youtube.com/watch?v=eYbvjPdQm24.

14. Rubén López-Bueno et al., "Global Prevalence of Cardiovascular Risk Factors Based on the Life's Essential 8 Score: An Overview of Systematic Reviews

and Meta-Analysis," *Cardiovascular Research* 120, no. 1 (February 27, 2024): 13–33, https://doi.org/10.1093/cvr/cvad176; Luigi Ferini-Strambi et al., "Role of Sleep in Neurodegeneration: The Consensus Report of the 5th Think Tank World Sleep Forum," *Neurological Sciences* 45, no. 2 (February 2024): 749–67, https://doi.org/10.1007/s10072-023-07232-7.

15. Joanna E. Wrede et al., "Mitochondrial DNA Copy Number in Sleep Duration Discordant Monozygotic Twins," *Sleep* 38, no. 10 (October 1, 2015): 1655–58, https://doi.org/10.5665/sleep.5068.

16. "Effect of Sleep Quality on Mitochondrial DNA Copy Number in Eveningness Chronotypes," *Journal of the College of Physicians and Surgeons Pakistan* 34, no. 01 (January 1, 2024): 73–77, https://doi.org/10.29271/jcpsp.2024.01.73.

17. Seolbin Han et al., "Association of Sleep Quality and Mitochondrial DNA Copy Number in Healthy Middle-Aged Adults," *Sleep Medicine* 113 (January 2024): 19–24, https://doi.org/10.1016/j.sleep.2023.11.011.

18. Nicholas J. Saner et al., "Exercise Mitigates Sleep-Loss-Induced Changes in Glucose Tolerance, Mitochondrial Function, Sarcoplasmic Protein Synthesis, and Diurnal Rhythms," *Molecular Metabolism* 43 (January 2021): 101110, https://doi.org/10.1016/j.molmet.2020.101110.

19. Tracey L. Sletten et al., "The Importance of Sleep Regularity: A Consensus Statement of the National Sleep Foundation Sleep Timing and Variability Panel," *Sleep Health* 9, no. 6 (December 2023): 801–20, https://doi.org/10.1016/j.sleh.2023.07.016.

20. "Screen Time and Sleep—It's Different for Adults | Restorative Sleep," *Lifestyle Medicine* (blog), August 8, 2024, https://longevity.stanford.edu/lifestyle/2024/08/08/screen-time-and-sleep-its-different-for-adults/.

21. Lauren E. Hartstein et al., "The Impact of Screen Use on Sleep Health across the Lifespan: A National Sleep Foundation Consensus Statement," *Sleep Health* 10, no. 4 (August 2024): 373–84, https://doi.org/10.1016/j.sleh.2024.05.001.

22. Ylva Hellsten et al., "Adenosine Concentrations in the Interstitium of Resting and Contracting Human Skeletal Muscle," *Circulation* 98, no. 1 (July 7, 1998): 6–8, https://doi.org/10.1161/01.CIR.98.1.6.

23. Brett A. Dolezal et al., "Interrelationship Between Sleep and Exercise: A Systematic Review," *Advances in Preventive Medicine* (March 16, 2017): 5979510, https://doi.org/10.1155/2017/1364387.

24. Do-Young Kim et al., "Systematic Review for the Medical Applications of Meditation in Randomized Controlled Trials," *International Journal of Environmental Research and Public Health* 19, no. 3 (January 22, 2022): 1244, https://doi.org/10.3390/ijerph19031244.

25. Uthman Albakri, Elizabeth Drotos, and Ree Meertens, "Sleep Health Promotion Interventions and Their Effectiveness: An Umbrella Review," *International Journal of Environmental Research and Public Health* 18, no. 11 (May 21, 2021): 5533, https://doi.org/10.3390/ijerph18115533.

Chapter 10:
Avoiding Prolonged Stress

1. "Chronic Stress Puts Your Health at Risk," Mayo Clinic, accessed December 13, 2024, https://www.mayoclinic.org/healthy-lifestyle/stress-management/in-depth/stress/art-20046037; Irini Manoli et al., "Mitochondria as Key Components of the Stress Response," *Trends in Endocrinology & Metabolism* 18, no. 5 (July 2007): 190–98, https://doi.org/10.1016/j.tem.2007.04.004.

2. Bruce S. McEwen, "Protective and Damaging Effects of Stress Mediators: Central Role of the Brain," *Dialogues in Clinical Neuroscience* 8, no. 4 (December 31, 2006): 367–81, https://doi.org/10.31887/DCNS.2006.8.4/bmcewen; Suzanne C. Segerstrom and Gregory E. Miller, "Psychological Stress and the Human Immune System: A Meta-Analytic Study of 30 Years of Inquiry," *Psychological Bulletin* 130, no. 4 (2004): 601–30, https://doi.org/10.1037/0033-2909.130.4.601.

3. Martin Picard et al., "An Energetic View of Stress: Focus on Mitochondria," *Frontiers in Neuroendocrinology* 49 (April 2018): 72–85, https://doi.org/10.1016/j.yfrne.2018.01.001; Natalia Bobba-Alves, Robert-Paul Juster, and Martin Picard, "The Energetic Cost of Allostasis and Allostatic Load," *Psychoneuroendocrinology* 146 (December 2022): 105951, https://doi.org/10.1016/j.psyneuen.2022.105951; Gee Euhn Choi and Ho Jae Han, "Glucocorticoid Impairs Mitochondrial Quality Control in Neurons," *Neurobiology of Disease* 152 (May 2021): 105301, https://doi.org/10.1016/j.nbd.2021.105301.

4. Yu Ruan et al., "Chronic Stress Hinders Sensory Axon Regeneration via Impairing Mitochondrial Cristae and OXPHOS," *Science Advances* 9, no. 40 (October 6, 2023): eadh0183, https://doi.org/10.1126/sciadv.adh0183.

5. J. K. Kiecolt-Glaser et al., "Slowing of Wound Healing by Psychological Stress," *The Lancet* 346, no. 8984 (November 1995): 1194–96, https://doi.org/10.1016/S0140-6736(95)92899-5.

6. Elissa S. Epel et al., "Wandering Minds and Aging Cells," *Clinical Psychological Science* 1, no. 1 (January 2013): 75–83, https://doi.org/10.1177/2167702612460234.

7. Elissa S. Epel et al., "Accelerated Telomere Shortening in Response to Life Stress," *Proceedings of the National Academy of Sciences* 101, no. 49 (December 7, 2004): 17312–15, https://doi.org/10.1073/pnas.0407162101.

8. Martin Picard et al., "A Mitochondrial Health Index Sensitive to Mood and Caregiving Stress," *Biological Psychiatry* 84, no. 1 (July 2018): 9–17, https://doi.org/10.1016/j.biopsych.2018.01.012.

9. Robert M. Sapolsky, *Why Zebras Don't Get Ulcers*, Third edition (New York: St. Martin's Griffin, 2004).

10. Fatih Ozbay et al., "Social Support and Resilience to Stress: From Neurobiology to Clinical Practice," *Psychiatry* 4, no. 5 (May 2007): 35–40.

11. Reena Debray et al., "Social Affiliation Predicts Mitochondrial DNA Copy Number in Female Rhesus Macaques," *Biology Letters* 15, no. 1 (January 2019): 20180643, https://doi.org/10.1098/rsbl.2018.0643.

12. Caroline Trumpff et al., "Psychosocial Experiences Are Associated with Human Brain Mitochondrial Biology," *Proceedings of the National Academy of Sciences* 121, no. 27 (July 2, 2024): e2317673121, https://doi.org/10.1073/pnas.2317673121.

13. PV AshaRani et al., "Purpose in Life in Older Adults: A Systematic Review on Conceptualization, Measures, and Determinants," *International Journal of Environmental Research and Public Health* 19, no. 10 (May 11, 2022): 5860, https://doi.org/10.3390/ijerph19105860.

14. Minhal Ahmed, Ivo Cerda, and Molly Maloof, "Breaking the Vicious Cycle: The Interplay Between Loneliness, Metabolic Illness, and Mental Health," *Frontiers in Psychiatry* 14 (March 8, 2023): 1134865, https://doi.org/10.3389/fpsyt.2023.1134865.

15. Joel D. Levine, "Chronically Lonely Flies Overeat and Lose Sleep," *Nature* 597, no. 7875 (September 9, 2021): 179–80, https://doi.org/10.1038/d41586-021-02194-2.

16. Andrew Steptoe et al., "Loneliness and Neuroendocrine, Cardiovascular, and Inflammatory Stress Responses in Middle-Aged Men and Women," *Psychoneuroendocrinology* 29, no. 5 (June 2004): 593–611, https://doi.org/10.1016/S0306-4530(03)00086-6; Leah D. Doane and Emma K. Adam, "Loneliness and Cortisol: Momentary, Day-to-Day, and Trait Associations," *Psychoneuroendocrinology* 35, no. 3 (April 2010): 430–41, https://doi.org/10.1016/j.psyneuen.2009.08.005; Sarah D. Pressman et al., "Loneliness, Social Network Size, and Immune Response to Influenza Vaccination in College Freshmen," *Health Psychology* 24, no. 3 (2005): 297–306, https://doi.org/10.1037/0278-6133.24.3.297.

17. Judith E. Carroll et al., "Negative Affective Responses to a Speech Task Predict Changes in Interleukin (IL)-6," *Brain, Behavior, and Immunity* 25, no. 2 (February 2011): 232–38, https://doi.org/10.1016/j.bbi.2010.09.024.

18. Caroline Trumpff et al., "Acute Psychological Stress Increases Serum Circulating Cell-Free Mitochondrial DNA," *Psychoneuroendocrinology* 106 (August 2019): 268–76, https://doi.org/10.1016/j.psyneuen.2019.03.026.

19. Fiona Hollis et al., "Neuroinflammation and Mitochondrial Dysfunction Link Social Stress to Depression," in *Neuroscience of Social Stress*, ed. Klaus A. Miczek and Rajita Sinha, vol. 54, Current Topics in Behavioral Neurosciences (Cham: Springer International Publishing, 2022), 59–93, https://doi.org/10.1007/7854_2021_300.

20. Kaizheng Duan et al., "Mitophagy in the Basolateral Amygdala Mediates Increased Anxiety Induced by Aversive Social Experience," *Neuron* 109, no. 23 (December 2021): 3793–3809.e8, https://doi.org/10.1016/j.neuron.2021.09.008; Lanmin Guo et al., "Repeated Social Defeat Stress Inhibits Development of Hippocampus Neurons Through Mitophagy and Autophagy," *Brain Research Bulletin* 182 (May 2022): 111–17, https://doi.org/10.1016/j.brainresbull.2022.01.009; Doğukan Hazar Ülgen, Silvie Rosalie Ruigrok, and Carmen Sandi, "Powering the Social Brain: Mitochondria in Social Behaviour," *Current Opinion in Neurobiology* 79 (April 2023): 102675, https://doi.org/10.1016/j.conb.2022.102675.

21. R. Takizawa et al., "Bullying Victimization in Childhood Predicts Inflammation and Obesity at Mid-Life: A Five-Decade Birth Cohort Study," *Psychological Medicine* 45, no. 13 (October 2015): 2705–15, https://doi.org/10.1017/S0033291715000653.

22. Fiona Hollis et al., "Mitochondrial Function in the Brain Links Anxiety with Social Subordination," *Proceedings of the National Academy of Sciences* 112, no. 50 (December 15, 2015): 15486–91, https://doi.org/10.1073/pnas.1512653112; Elias Gebara et al., "Mitofusin-2 in the Nucleus Accumbens Regulates Anxiety and Depression-like Behaviors Through Mitochondrial and Neuronal Actions," *Biological Psychiatry* 89, no. 11 (June 2021): 1033–44, https://doi.org/10.1016/j.biopsych.2020.12.003.

23. Lorenz Goette et al., "Stress Pulls Us Apart: Anxiety Leads to Differences in Competitive Confidence Under Stress," *Psychoneuroendocrinology* 54 (April 2015): 115–23, https://doi.org/10.1016/j.psyneuen.2015.01.019.

24. "Chronic Stress Puts Your Health at Risk," Mayo Clinic, accessed December 13, 2024, https://www.mayoclinic.org/healthy-lifestyle/stress-management/in-depth/stress/art-20046037.

25. "Amid Loneliness Epidemic, Surgeon General Asks Students to Focus on Connections | ASU News," accessed April 23, 2025, https://news.asu.edu/20231114-sun-devil-life-amid-loneliness-epidemic-surgeon-general

-asks-students-focus-connections; Alexa Mikhail, "Try This 5-for-5 Challenge to Combat Loneliness: 'It's Incredibly Powerful,'" Fortune Well, accessed April 23, 2025, https://fortune.com/well/article/loneliness-5-for-5-challenge/.

26. Manoj K. Bhasin et al., "Relaxation Response Induces Temporal Transcriptome Changes in Energy Metabolism, Insulin Secretion and Inflammatory Pathways," ed. Yidong Bai, *PLOS One* 8, no. 5 (May 1, 2013): e62817, https://doi.org/10.1371/journal.pone.0062817.

27. Alexandra D. Crosswell et al., "Deep Rest: An Integrative Model of How Contemplative Practices Combat Stress and Enhance the Body's Restorative Capacity," *Psychological Review* 131, no. 1 (January 2024): 247–70, https://doi.org/10.1037/rev0000453; Sarah Karrasch et al., "Randomized Controlled Trial Investigating Potential Effects of Relaxation on Mitochondrial Function in Immune Cells: A Pilot Experiment," *Biological Psychology* 183 (October 2023): 108656, https://doi.org/10.1016/j.biopsycho.2023.108656.

28. Gandhar V. Mandlik et al., "Effect of a Single Session of Yoga and Meditation on Stress Reactivity: A Systematic Review," *Stress and Health* 40, no. 3 (June 2024): e3324, https://doi.org/10.1002/smi.3324.

29. Michele Antonelli, Grazia Barbieri, and Davide Donelli, "Effects of Forest Bathing (Shinrin-Yoku) on Levels of Cortisol as a Stress Biomarker: A Systematic Review and Meta-Analysis," *International Journal of Biometeorology* 63, no. 8 (August 2019): 1117–34, https://doi.org/10.1007/s00484-019-01717-x.

30. MaryCarol R. Hunter, Brenda W. Gillespie, and Sophie Yu-Pu Chen, "Urban Nature Experiences Reduce Stress in the Context of Daily Life Based on Salivary Biomarkers," *Frontiers in Psychology* 10 (April 4, 2019): 722, https://doi.org/10.3389/fpsyg.2019.00722.

31. Mathew P. White et al., "Spending at Least 120 Minutes a Week in Nature Is Associated with Good Health and Wellbeing," *Scientific Reports* 9, no. 1 (June 13, 2019): 7730, https://doi.org/10.1038/s41598-019-44097-3.

32. Caroline Kaercher Kramer and Cristiane Bauermann Leitao, "Laughter as Medicine: A Systematic Review and Meta-Analysis of Interventional Studies Evaluating the Impact of Spontaneous Laughter on Cortisol Levels," ed. Fares Alahdab, *PLOS One* 18, no. 5 (May 23, 2023): e0286260, https://doi.org/10.1371/journal.pone.0286260.

33. Lee S. Berk, Stanley A. Tan, and Dottie Berk, "Cortisol and Catecholamine Stress Hormone Decrease Is Associated with the Behavior of Perceptual Anticipation of Mirthful Laughter," *The FASEB Journal* 22, no. S1 (March 2008), https://doi.org/10.1096/fasebj.22.1_supplement.946.11.

34. Robert M. Sapolsky, *Why Zebras Don't Get Ulcers*, Third edition (New York: St. Martin's Griffin, 2004); Jay R. Kaplan, Haiying Chen, and Stephen B. Manuck, "The Relationship Between Social Status and Atherosclerosis in Male and Female Monkeys as Revealed by Meta-Analysis," *American Journal of Primatology* 71, no. 9 (September 2009): 732–41, https://doi.org/10.1002/ajp.20707.

35. Juan M. Suárez-Rivero et al., "From Mitochondria to Atherosclerosis: The Inflammation Path," *Biomedicines* 9, no. 3 (March 5, 2021): 258, https://doi.org/10.3390/biomedicines9030258.

36. Elissa Epel, *The Stress Prescription: Seven Days to More Joy and Ease* (New York: Penguin Books, 2022).

37. Stefan G. Hofmann et al., "The Efficacy of Cognitive Behavioral Therapy: A Review of Meta-Analyses," *Cognitive Therapy and Research* 36, no. 5 (October 2012): 427–40, https://doi.org/10.1007/s10608-012-9476-1.

38. Natalia Bobba-Alves, Robert-Paul Juster, and Martin Picard, "The Energetic Cost of Allostasis and Allostatic Load," *Psychoneuroendocrinology* 146 (December 2022): 105951, https://doi.org/10.1016/j.psyneuen.2022.105951.

39. Alexandra D. Crosswell et al., "Deep Rest: An Integrative Model of How Contemplative Practices Combat Stress and Enhance the Body's Restorative Capacity," *Psychological Review* 131, no. 1 (January 2024): 247–70, https://doi.org/10.1037/rev0000453.

Chapter 11:
Avoid These to Help Your Mitochondria

1. Center for Health Security, "Cyanide," 2022, https://centerforhealthsecurity.org/sites/default/files/2023-02/cyanide.pdf.

2. Rui Guo and Jun Ren, "Alcohol and Acetaldehyde in Public Health: From Marvel to Menace," *International Journal of Environmental Research and Public Health* 7, no. 4 (March 25, 2010): 1285–1301, https://doi.org/10.3390/ijerph7041285; Che-Hong Chen et al., "Activation of Aldehyde Dehydrogenase-2 Reduces Ischemic Damage to the Heart," *Science* 321, no. 5895 (September 12, 2008): 1493–95, https://doi.org/10.1126/science.1158554.

3. P. E. Ronksley et al., "Association of Alcohol Consumption with Selected Cardiovascular Disease Outcomes: A Systematic Review and Meta-Analysis," *BMJ* 342, no. feb22 1 (February 22, 2011): d671–d671, https://doi.org/10.1136/bmj.d671.

4. Che-Hong Chen et al., "Activation of Aldehyde Dehydrogenase-2 Reduces Ischemic Damage to the Heart," *Science* 321, no. 5895 (September 12, 2008): 1493–95, https://doi.org/10.1126/science.1158554.

5. "No Level of Alcohol Consumption Is Safe for Our Health," accessed December 13, 2024, https://www.who.int/europe/news-room/04-01-2023-no-level-of-alcohol-consumption-is-safe-for-our-health.

6. "Even Just 1 Alcoholic Drink a Day May Increase Blood Pressure," www.heart.org, accessed December 13, 2024, https://www.heart.org/en/news/2023/07/31/even-just-1-alcoholic-drink-a-day-may-increase-blood-pressure; American Cancer Society, "Alcohol Use and Cancer," accessed December 13, 2024, https://www.cancer.org/cancer/risk-prevention/diet-physical-activity/alcohol-use-and-cancer.html.

7. Roni Caryn Rabin, "Surgeon General Calls for Cancer Warnings on Alcohol," *New York Times*, January 3, 2025, https://www.nytimes.com/2025/01/03/health/alcohol-surgeon-general-warning.html.

8. Ville Salaspuro and Mikko Salaspuro, "Synergistic Effect of Alcohol Drinking and Smoking on *in Vivo* Acetaldehyde Concentration in Saliva," *International Journal of Cancer* 111, no. 4 (September 10, 2004): 480–83, https://doi.org/10.1002/ijc.20293.

9. Sergey Dikalov et al., "Tobacco Smoking Induces Cardiovascular Mitochondrial Oxidative Stress, Promotes Endothelial Dysfunction, and Enhances Hypertension," *American Journal of Physiology-Heart and Circulatory Physiology* 316, no. 3 (March 1, 2019): H639–46, https://doi.org/10.1152/ajpheart.00595.2018.

10. U.S. Food & Drug Administration, "Harmful and Potentially Harmful Constituents in Tobacco Products and Tobacco Smoke: Established List," *FDA*, July 21, 2020, https://www.fda.gov/tobacco-products/rules-regulations-and-guidance-related-tobacco-products/harmful-and-potentially-harmful-constituents-tobacco-products-and-tobacco-smoke-established-list; **Formaldehyde:** Christy B. M. Tulen et al., "Disruption of the Molecular Regulation of Mitochondrial Metabolism in Airway and Lung Epithelial Cells by Cigarette Smoke: Are Aldehydes the Culprit?," *Cells* 12, no. 2 (January 12, 2023): 299, https://doi.org/10.3390/cells12020299. **Arsenic:** Nathan Earl Rainey, Anne-Sophie Armand, and Patrice X. Petit, "Sodium Arsenite and Arsenic Trioxide Differently Affect the Oxidative Stress of Lymphoblastoid Cells: An Intricate Crosstalk Between Mitochondria, Autophagy and Cell Death," ed. Prashanta Kumar Panda, *PLOS One* 19, no. 5 (May 10, 2024): e0302701, https://doi.org/10.1371/journal.pone.0302701; Chandra

Prakash, Sunil Chhikara, and Vijay Kumar, "Mitochondrial Dysfunction in Arsenic-Induced Hepatotoxicity: Pathogenic and Therapeutic Implications," *Biological Trace Element Research* 200, no. 1 (January 2022): 261–70, https://doi.org/10.1007/s12011-021-02624-2.

11. Jessica L. Fetterman, Melissa J. Sammy, and Scott W. Ballinger, "Mitochondrial Toxicity of Tobacco Smoke and Air Pollution," *Toxicology* 391 (November 2017): 18–33, https://doi.org/10.1016/j.tox.2017.08.002.

12. Che-Hong Chen et al., "Activation of Aldehyde Dehydrogenase-2 Reduces Ischemic Damage to the Heart," *Science* 321, no. 5895 (September 12, 2008): 1493–95, https://doi.org/10.1126/science.1158554.

13. Mumiye A. Ogunwale et al., "Aldehyde Detection in Electronic Cigarette Aerosols," *ACS Omega* 2, no. 3 (March 31, 2017): 1207–14, https://doi.org/10.1021/acsomega.6b00489.

14. Laura LeBlanc, "Vaping—Not Prior Smoking—Is Associated with Changes in Gene Regulation Linked to Disease," Newsroom, November 23, 2021, https://keck.usc.edu/news/vaping-not-prior-smoking-is-associated-with-changes-in-gene-regulation-linked-to-disease/.

15. Jieliang Li et al., "Electronic Cigarettes Induce Mitochondrial DNA Damage and Trigger TLR9 (Toll-Like Receptor 9)-Mediated Atherosclerosis," *Arteriosclerosis, Thrombosis, and Vascular Biology* 41, no. 2 (February 2021): 839–53, https://doi.org/10.1161/ATVBAHA.120.315556.

16. "Vaping's Respiratory Effects Traced by Leading Basic Researcher," National Institute of Environmental Health Sciences, accessed December 13, 2024, https://factor.niehs.nih.gov/2023/4/feature/4-e-cigarette-research.

17. Center for Disease Control, "Polycyclic Aromatic Hydrocarbons (PAHs)," 2009, https://www.epa.gov/sites/default/files/2014-03/documents/pahs_factsheet_cdc_2013.pdf.

18. "Five Steps for Cancer-Safe Grilling," *American Institute for Cancer Research* (blog), accessed December 13, 2024, https://www.aicr.org/news/five-steps-for-cancer-safe-grilling/; "A Healthy Summer Cookout," *The Nutrition Source* (blog), June 29, 2017, https://nutritionsource.hsph.harvard.edu/2017/06/29/healthy-summer-picnic/.

19. AGE in specific foods: Jaime Uribarri et al., "Advanced Glycation End Products in Foods and a Practical Guide to Their Reduction in the Diet," *Journal of the American Dietetic Association* 110, no. 6 (June 2010): 911–916.e12, https://doi.org/10.1016/j.jada.2010.03.018.

20. Yusuke Hirata et al., "Trans-Fatty Acids Facilitate DNA Damage-Induced Apoptosis Through the Mitochondrial JNK-Sab-ROS Positive Feedback Loop,"

Scientific Reports 10, no. 1 (February 17, 2020): 2743, https://doi.org/10.1038/s41598-020-59636-6; Rafael Longhi et al., "Effect of a Trans Fatty Acid-Enriched Diet on Mitochondrial, Inflammatory, and Oxidative Stress Parameters in the Cortex and Hippocampus of Wistar Rats," *European Journal of Nutrition* 57, no. 5 (August 2018): 1913–24, https://doi.org/10.1007/s00394-017-1474-3.

21. "Shining the Spotlight on Trans Fats," *The Nutrition Source* (blog), September 18, 2012, https://nutritionsource.hsph.harvard.edu/what-should-you-eat/fats-and-cholesterol/types-of-fat/transfats/; "Yes, Trans Fats Still Exist After the Ban," Cleveland Clinic, accessed December 13, 2024, https://health.cleveland clinic.org/why-trans-fats-are-bad-for-you.

22. "Trans Fat: Double Trouble for Your Heart," Mayo Clinic, accessed December 13, 2024, https://www.mayoclinic.org/diseases-conditions/high-blood-cholesterol/in-depth/trans-fat/art-20046114.

23. Pawel Hikisz and Damian Jacenik, "Diet as a Source of Acrolein: Molecular Basis of Aldehyde Biological Activity in Diabetes and Digestive System Diseases," *International Journal of Molecular Sciences* 24, no. 7 (March 31, 2023): 6579, https://doi.org/10.3390/ijms24076579.

24. Tetsumori Yamashima et al., "Vegetable Oil-Peroxidation Product 'Hydroxynonenal' Causes Hepatocyte Injury and Steatosis via Hsp70.1 and BHMT Disorders in the Monkey Liver," *Nutrients* 15, no. 8 (April 14, 2023): 1904, https://doi.org/10.3390/nu15081904.

25. Jan F. Stevens and Claudia S. Maier, "Acrolein: Sources, Metabolism, and Biomolecular Interactions Relevant to Human Health and Disease," *Molecular Nutrition & Food Research* 52, no. 1 (January 2008): 7–25, https://doi.org/10.1002/mnfr.200700412; Pawel Hikisz and Damian Jacenik, "Diet as a Source of Acrolein: Molecular Basis of Aldehyde Biological Activity in Diabetes and Digestive System Diseases," *International Journal of Molecular Sciences* 24, no. 7 (March 31, 2023): 6579, https://doi.org/10.3390/ijms24076579.

26. Hongxia Liao, Mengting Zhu, and Yi Chen, "4-Hydroxy-2-Nonenal in Food Products: A Review of the Toxicity, Occurrence, Mitigation Strategies and Analysis Methods," *Trends in Food Science & Technology* 96 (February 2020): 188–98, https://doi.org/10.1016/j.tifs.2019.12.011.

27. Tetsumori Yamashima, *Cooking Oil Makes Genius or Alzheimer!* (London: Austin Macauley Publishers, 2017), 8.

28. Tetsumori Yamashima, *Cooking Oil Makes Genius or Alzheimer!* (London: Austin Macauley Publishers, 2017), 99.

29. Martha S. Sandy et al., "Role of Active Oxygen in Paraquat and 1-Methyl-4-Phenyl-1,2,3,6-Tetrahydropyridine (MPTP) Cytotoxicity," in *Oxygen Radi-*

cals in Biology and Medicine, ed. Michael G. Simic et al. (Boston: Springer US, 1988), 795–801, https://doi.org/10.1007/978-1-4684-5568-7_127; J. William Langston, "The MPTP Story," *Journal of Parkinson's Disease* 7, no. s1 (March 6, 2017): S11–19, https://doi.org/10.3233/JPD-179006.

30. Bret Stetka, "Parkinson's Disease and Pesticides: What's the Connection?," *Scientific American*, accessed December 13, 2024, https://www.scientificameri can.com/article/parkinsons-disease-and-pesticides-whats-the-connection/.

31. S. Costello et al., "Parkinson's Disease and Residential Exposure to Maneb and Paraquat from Agricultural Applications in the Central Valley of California," *American Journal of Epidemiology* 169, no. 8 (March 3, 2009): 919–26, https://doi.org/10.1093/aje/kwp006.

32. "How Can I Wash Pesticides from Fruit and Veggies?" accessed December 14, 2024, https://npic.orst.edu/faq/fruitwash.html.

33. Caroline M. Tanner et al., "Rotenone, Paraquat, and Parkinson's Disease," *Environmental Health Perspectives* 119, no. 6 (June 2011): 866–72, https://doi.org/10.1289/ehp.1002839.

34. "The Rise of Parkinson's Disease," *American Scientist*, March 27, 2020, https://www.americanscientist.org/article/the-rise-of-parkinsons-disease.

35. "Low Level Exposure to Air Pollution Is Harmful, Mouse Model Shows," EurekAlert!, accessed December 13, 2024, https://www.eurekalert.org/news-releases/823782.

36. Fox News, "Oklahoma Mom Warns of Carbon Monoxide Poisoning After Son, 9, Dies on Boating Trip," accessed December 13, 2024, https://www.foxnews.com/health/oklahoma-mom-carbon-monoxide-poisoning-son-death-boating-trip; "'Andy's Law' Passes House; Would Require Boat Carbon Monoxide Stickers," Oklahoma House of Representatives, accessed March 27, 2025, https://www.okhouse.gov/posts/news-20230308_1.

37. Jason J. Rose et al., "Carbon Monoxide Poisoning: Pathogenesis, Management, and Future Directions of Therapy," *American Journal of Respiratory and Critical Care Medicine* 195, no. 5 (March 1, 2017): 596–606, https://doi.org/10.1164/rccm.201606-1275CI.

38. CDC, "Carbon Monoxide Poisoning Basics," Carbon Monoxide Poisoning, October 31, 2024, https://www.cdc.gov/carbon-monoxide/about/index.html.

39. Source: Johns Hopkins "Carbon Monoxide Poisoning," August 20, 2024, https://www.hopkinsmedicine.org/health/conditions-and-diseases/carbon-monoxide-poisoning; CDC, "About CO Poisoning on Your Boat," Carbon Monoxide Poisoning, April 21, 2024, https://www.cdc.gov/carbon-monoxide/about/boating.html.

40. "Ultraviolet Waves—NASA Science," accessed December 13, 2024, https://science.nasa.gov/ems/10_ultravioletwaves/.

41. Annapoorna Sreedhar, Leopoldo Aguilera-Aguirre, and Keshav K. Singh, "Mitochondria in Skin Health, Aging, and Disease," *Cell Death & Disease* 11, no. 6 (June 9, 2020): 444, https://doi.org/10.1038/s41419-020-2649-z.

42. Mark Berneburg et al., "Chronically Ultraviolet-Exposed Human Skin Shows a Higher Mutation Frequency of Mitochondrial DNA as Compared to Unexposed Skin and the Hematopoietic System," *Photochemistry and Photobiology* 66, no. 2 (August 1997): 271–75, https://doi.org/10.1111/j.1751-1097.1997.tb08654.x.

43. Mark Berneburg et al., "Induction of the Photoaging-Associated Mitochondrial Common Deletion *In Vivo* in Normal Human Skin," *Journal of Investigative Dermatology* 122, no. 5 (May 2004): 1277–83, https://doi.org/10.1111/j.0022-202X.2004.22502.x.

44. Brendan T. Heiden et al., "Assessment of Formal Tobacco Treatment and Smoking Cessation in Dual Users of Cigarettes and E-Cigarettes," *Thorax* 78, no. 3 (March 2023): 267–73, https://doi.org/10.1136/thorax-2022-218680.

45. John F. Kelly, Keith Humphreys, and Marica Ferri, "Alcoholics Anonymous and Other 12-Step Programs for Alcohol Use Disorder," ed. Cochrane Drugs and Alcohol Group, *Cochrane Database of Systematic Reviews*, March 11, 2020, https://doi.org/10.1002/14651858.CD012880.pub2.

46. National Safety Council, "'If You Don't Drink, Don't Start,' Heart Researchers Say," *Family Safety & Health*, June 22, 2022, accessed December 13, 2024, https://www.safetyandhealthmagazine.com/articles/22733-if-you-dont-drink-dont-start-heart-researchers-say.

47. OCSPP US EPA, "Learn About Pollution Prevention," Overviews and Factsheets, June 4, 2013, https://www.epa.gov/p2/learn-about-pollution-prevention.

48. Valery Lemmens et al., "Effectiveness of Smoking Cessation Interventions Among Adults: A Systematic Review of Reviews," *European Journal of Cancer Prevention* 17, no. 6 (November 2008): 535–44, https://doi.org/10.1097/CEJ.0b013e3282f75e48.

49. "Read the Surgeon General's 1964 Report on Smoking and Health," PBS News, January 12, 2014, https://www.pbs.org/newshour/health/first-surgeon-general-report-on-smokings-health-effects-marks-50-year-anniversary.

50. Ted Alcorn, "Should Alcoholic Beverages Have Cancer Warning Labels?," *New York Times*, April 9, 2024, https://www.nytimes.com/2024/04/09/health/alcohol-cancer-warning.html.

Epilogue:

1. Paolo Bernardi, Michela Carraro, and Giovanna Lippe, "The Mitochondrial Permeability Transition: Recent Progress and Open Questions," *The FEBS Journal* 289, no. 22 (November 2022): 7051–74, https://doi.org/10.1111/febs.16254.

2. Luca Ermini et al., "Complete Mitochondrial Genome Sequence of the Tyrolean Iceman," *Current Biology* 18, no. 21 (November 2008): 1687–93, https://doi.org/10.1016/j.cub.2008.09.028.

3. V. Coia et al., "Whole Mitochondrial DNA Sequencing in Alpine Populations and the Genetic History of the Neolithic Tyrolean Iceman," *Scientific Reports* 6, no. 1 (January 14, 2016): 18932, https://doi.org/10.1038/srep18932.

4. Niall J. O'Sullivan et al., "A Whole Mitochondria Analysis of the Tyrolean Iceman's Leather Provides Insights into the Animal Sources of Copper Age Clothing," *Scientific Reports* 6, no. 1 (August 18, 2016): 31279, https://doi.org/10.1038/srep31279; https://www.science.org/content/article/tzi-iceman-had-some-wild-clothes.

5. "Mitochondria Make a Comeback," *Science* 283, no. 5407 (March 5, 1999): 1475, https://doi.org/10.1126/science.283.5407.1475.

INDEX

ABOUT THE AUTHORS

Daria Mochly-Rosen is the George D. Smith Professor for Translational Medicine at Stanford University School of Medicine. She's passionate about translating scientific discoveries into treatments and demystifying science to help people make educated health decisions. Based on her discoveries, Daria has founded three biotech companies. She has also founded SPARK, a program that focuses on translating scientific innovations to benefit patients. Following the success of the program at Stanford, leading to the formation of more than fifty biopharma companies, SPARK now operates in dozens of universities around the world.

Emanuel Rosen is a bestselling author whose books have been translated into thirteen languages. Throughout his writing career, he's been focusing on simplifying complex concepts by using plain language and storytelling. He's the author of the national bestseller *The Anatomy of Buzz*, and the award-winning *Absolute Value* (with Stanford professor Itamar Simonson). Emanuel was previously vice president of marketing at Niles Software, where he launched the company's flagship product, EndNote. He's presented his work in numerous forums, including at companies such as Google, Intel, and Nike.

Daria and Emanuel are married, live in Menlo Park, California, and have four adult children and three grandchildren.